Contributions to

Scalable Wireless ad hoc Networks

Supporting Quality-of-Service

New Concepts for Radio Resource Management, Switching,
Transmission in Sensor Networks, Integrated R&D Platform

vom Fachbereich Elektrotechnik, Informationstechnik, Medientechnik
der Bergischen Universität Wuppertal
genehmigte

Dissertation

---

zur Erlangung des akademischen Grades
eines Doktor-Ingenieurs

von
Dipl.-Ing. Arasch Honarbacht
Teheran

Wuppertal 2004

Tag der mündlichen Prüfung: 17.12.2004
Hauptreferent: Prof. Dr.-Ing. Anton Kummert
Korreferent: Prof. Dr.-Ing. Heinz Chaloupka

Bibliographic information published by Die Deutsche Bibliothek

Die Deutsche Bibliothek lists this publication in the Deutsche Nationalbibliografie; detailed bibliographic data is available in the Internet at http://dnb.ddb.de.

ISBN 3-8325-0842-2

Logos Verlag Berlin
Comeniushof, Gubener Str. 47,
10243 Berlin
Tel.: +49 030 42 85 10 90
Fax: +49 030 42 85 10 92
INTERNET: http://www.logos-verlag.de

Vorwort

Die vorliegende Arbeit entstand während meiner Tätigkeit als wissenschaftlicher Mitarbeiter am Lehrstuhl für Allgemeine Elektrotechnik und Theoretische Nachrichtentechnik der Bergischen Universität Wuppertal.

Mein besonderer Dank gilt Herrn Prof. Dr.-Ing. Anton Kummert für die Anregung und Betreuung dieser Arbeit. Herrn Prof. Dr.-Ing. Heinz Chaloupka danke ich für die Bereitschaft zur Übernahme des Korreferats und für sein dieser Arbeit entgegengebrachtes Interesse.

Ebenso danke ich allen Kollegen und Studenten, die mich durch ihren Einsatz auf verschiedenste Art bei der Durchführung meiner Untersuchungen unterstützt haben.

Düsseldorf, März 2005

This page intentionally left blank.

Contents

Part III   Demonstrator Design   141

Appendices

List of Figures

List of Tables

This page intentionally left blank.

Listings

This page intentionally left blank.

Abbreviations and Acronyms

2G/3G/4G(+)	second/third/fourth generation (and beyond)
3D	three-dimensional
4GMF	Fourth Generation Mobile Forum
ABR	available bit rate
ACK	acknowledgment
ADC	analog-to-digital converter
AFE	analog front-end
AGC	automatic gain control
ARP	Address Resolution Protocol
ARQ	automatic repeat request
ASIC	application specific integrated circuit
ASRAM	asynchronous static RAM
ATM	Asynchronous Transfer Mode
BER	bit error rate
BPSK	binary phase shift keying
BRAM	block RAM
BS	base station
BSS	basic service set
CBR	constant bit rate
CCA	clear channel assessment
CCK	complementary code keying
CDMA	code division multiple access
CIC	cascaded integrator comb
CINR	carrier-to-interference-plus-noise ratio
CMU	Carnegie Mellon University
CRC	cyclic redundancy check

CS	carrier sense
CSMA(/CA)	carrier sense multiple access (with collision avoidance)
CTS	clear-to-send
CW	contention window
DAC	digital-to-analog converter
DARPA	Defense Advanced Research Projects Agency
DBPSK	differential BPSK
DCA	dynamic channel assignment
DCF	Distributed Coordination Function
DDC	digital down-conversion
DE	discrete-event
DFE	digital front-end
DIFS	distributed (coordination function) interframe space
DLC	data link control
DMAC	dedicated channel medium access control
DMF	Delay Measurement Function
DoD	(United States) Department of Defense
DQPSK	differential QPSK
DSK	development starter kit
DSP[1]	digital signal processing
DSP[2]	digital signal processor
DSRC	Dedicated Short Range Communications
DSSS	direct sequence spread spectrum
DUC	digital up-conversion
EDGE	Enhanced Data Rates for Global Evolution
EIFS	extended interframe space
EIRP	effective isotropic radiated power
EMIF	external memory interface
FCA	fixed channel assignment
FDMA	frequency division multiple access
FEC	forward error correction
FHSS	frequency hopping spread spectrum
FIFO	first-in, first-out
FIR	finite impulse response
FPGA	field programmable gate array
FSM	finite state machine
GFSK	GAUSSian frequency shift keying
GMSK	GAUSSian minimum shift keying
GPP	general purpose processor

GPS	Global Positioning System
GSM	Global System for Mobile Communications
HAL	hardware abstraction layer
IBSS	independent basic service set
IEEE	Institute of Electrical and Electronics Engineers
IETF	Internet Engineering Task Force
I	in-phase ($\Re\{\cdot\}$ part)
IF	intermediate frequency
IFS	interframe space
ITU	International Telecommunication Union
IP(v4/v6)	Internet Protocol (Version 4/6)
JTAG	Joint Test Action Group
KB	kilo byte
kbps	kilo bit per second
kSps	kilo samples per second
L1/L2/.../L7	ISO/OSI layer 1/2/.../7
LAN	local area network
LLC	logical link control
LNA	low-noise amplifier
LO	local oscillator
LOS	line-of-sight
MAC[1]	medium access control
MAC[2]	multiply, accumulate
MAN	metropolitan area network
MANET	mobile ad hoc network
MB	mega byte
Mb	mega bit
Mbps	mega bit per second
McBSP	multichannel buffered serial port
MCps	mega chips per second
MIB	management information base
MIMO	multiple-input, multiple-output
MIT	Massachusetts Institute of Technology
mITF	Mobile IT Forum
MLME	MAC sublayer management entity
MMPDU	MAC management protocol data unit
MPDU	MAC protocol data unit
MPEG	Motion Picture Experts Group
MPLS	multiprotocol label switching

MSDU	MAC service data unit
MSps	mega samples per second
NAV	network allocation vector
NCO	numerically controlled oscillator
NDP	Neighborhood Discovery Protocol
NIC	network interface card
NLOS	non-line-of-sight
OFDM	orthogonal frequency division multiplexing
OFDMA	orthogonal frequency division multiple access
OVSF	orthogonal variable spreading factor (codes)
P2P	peer-to-peer
PAN	personal area network
PBCC	packet binary convolutional coding
PC	personal computer
PCB	printed circuit board
PCF	Point Coordination Function
PDA	personal digital assistant
PDU	protocol data unit
PHY	physical layer
PLCP	physical layer convergence protocol
PLME	physical layer management entity
PLR	packet loss ratio
PMD	physical medium dependent (device)
PSK	phase shift keying
PSTN	Public Switched Telephone Network
Q	quadrature ($\Im\{\cdot\}$ part)
QAM[1]	quadrature amplitude modulation
QAM[2]	quadrature amplitude mixer
QDMA	quadrature division multiple access
QoS	quality-of-service
QPSK	quaternary PSK
RAM	random access memory
RACH	random access channel
RCF	RAM coefficient filter
R&D	research & development
RF	radio frequency
RFC	request for comments
ROM	read-only memory
RERR	route error

RREP	route reply
RREQ	route request
RISC	reduced instruction set computer
RRM	radio resource management
RSP	receive signal processor
RSVP	Resource Reservation Protocol
RTS	request-to-send
RX	receive, receiver
SAW	surface acoustic wave
SBSRAM	synchronous burst static RAM
SC	single carrier
SDMA	spatial division multiple access
SDL	specification and description language
SDR	software defined radio
SDRAM	synchronous dynamic RAM
SDU	service data unit
SFDR	spurious-free dynamic range
SID	sensor identifier
SIFS	short interframe space
SME	station management entity
SNR	signal-to-noise ratio
STL	(C++) standard template library
SoC	system-on-chip
TBSD-RRM	transaction-based soft-decision RRM
TCP	Transmission Control Protocol
TDMA	time division multiple access
TDD	time division duplex
TSF	Timer Synchronization Function
TSP	transmit signal processor
TTL	time-to-live
TX	transmit, transmitter
UBR	unspecified bit rate
UDP	User Datagram Protocol
UMTS	Universal Mobile Telecommunications System
USB	Universal Serial Bus
VBR	variable bit rate
VGA	variable gain amplifier
VHDL	Very High Speed Integrated Circuit Hardware Description Language
VLIW	very long instruction word

WAN	wide area network
WCDMA	wide-band CDMA
WLAN	wireless local area network
WPAN	wireless personal area network
WM	wireless medium
WSDP	Wireless Sensor Data Protocol
WSN	wireless sensor network
WWANS	Wireless Wide Area Network Simulator
WWRF	Wireless World Research Forum

Abstract

Mobile ad hoc networks (also known as multihop or mesh networks) are attracting more and more interest from the research community. Originally designed for military applications (packet radio), the idea of spontaneous communication using direct links to immediate neighbors has rapidly spread over to the civilian sector. However, these are still the beginnings of ad hoc technology, since, up to now, technically feasible systems are usually confined to the body, personal or local area, and a single-hop topology. Moreover, some privileged station usually takes the role of a central coordinator that assigns time slots, performs timer synchronization and other centralized management tasks. Some networks may be operated without such a coordinator – but this comes at the expense of degraded performance in terms of throughput/delay characteristics.

The driving force behind this thesis is the idea of contributing to the design of a practicable wireless multihop network without any centralized management at all. Furthermore, the architecture and its building blocks shall be scalable, such that metropolitan or even wide area networks may be created based on the presented concepts with appropriate quality-of-service (QoS). For obvious reasons, it is not possible to cover all of the important aspects involved in the design of such a system. As a matter of fact, the thesis will start off with a novel architecture, setting the scene for the various building blocks, which are going to be described and evaluated in consecutive chapters. The proposed architecture allows contracting differentiated QoS classes (with respect to bandwidth requirements, delay and jitter) for integrated services. In particular, minimizing the end-to-end delay in a multihop scenario has been a primary design criterion.

Organization

Part I

In the first part of this thesis, a number of new ideas, algorithms and protocols for multihop wireless ad hoc networks are presented. An innovative radio resource management protocol is proposed that implements a fully distributed dynamic channel assignment algorithm for multihop networks. Thereby it eliminates the hidden and exposed terminal problems. During the development of this protocol, extensions to the frequently used IEEE 802.11 medium access control have evolved. Since they are also useful in other contexts, they are presented as a self-contained topic. Furthermore, wireless sensor networks are identified as a new field of application for classic KALMAN filtering. A ready-to-use protocol is introduced, letting sensor networks profit from this powerful tool of statistical signal processing.

Part II

The second part is dedicated to a novel software-based simulation environment: The Wireless Wide Area Network Simulator. It is applicable to a wide range of wireless ad hoc networks, spanning wireless sensor networks, single-hop wireless local area networks, and metropolitan or even wide area multihop ad hoc networks. This software has been used to verify and optimize the protocols designed in the first part. In addition to those protocols, an all-embracing suite of complementary protocols has been developed particularly for this simulator. Among other benefits, this concept provides for **system level** simulation of the proposed telecommunications network and cross-layer optimization approaches. Measures have been taken, which make the developed protocol stack ready to be employed in real-world devices, such as the hardware demonstrator outlined in the third part.

Part III

Finally, the design of a suitable hardware demonstrator is introduced. Often, research in multihop ad hoc networks is done by computer scientists with little knowledge of the underlying hardware structures. Most testbeds for multihop ad hoc networks used to date are based on network interface cards developed after the IEEE 802.11 standard, albeit multihop performance of this protocol is known to be poor. Therefore, a very flexible transceiver hardware based on state-of-the-art software defined radio technology has been designed, which integrates with the simulator. This hardware enables new ways of switching in wireless multihop networks, leading to minimal forwarding delays. It shall serve as a proof-of-concept, in that it demonstrates the technical feasibility of the proposed building blocks and overall system architecture.

## CHAPTER 1

Introduction

Covering fundamental preliminary topics, this chapter is intended as a review of some elementary concepts and building blocks, which will be referenced in subsequent chapters. Beginning with a survey of common wireless network architectures, their properties are briefly discussed. Following that, a few topics in quality-of-service are presented. Next, the ad hoc networking paradigm as such is elaborated on, including a short overview of history, drivers, advantages and challenges. Afterwards, basics of protocol design and the associated terminology are outlined. Principles of random access protocols, leading to the development of CSMA/CA, are provided. These concepts are particularly important for the comprehension of consecutive parts and chapters. Finally, some related research projects, standards and commercial products are concisely introduced.

1.1 Wireless Network Architectures

Three types of wireless network architectures can be distinguished: each type is best-suited for a different kind of application.

1.1.1 Point-to-Point Networks

A point-to-point network (figure 1.1) is composed of two radios sharing an immediate wireless link. These radios are usually operated in line-of-sight (LOS). Thus, they are well suited for high data rates: multipath phenomena and multiple access overheads do not strain throughput. Hence, they are often used for dedicated corporate connections (so-called "leased lines"), wireless

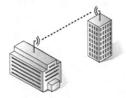

Figure 1.1: Point-to-Point Network

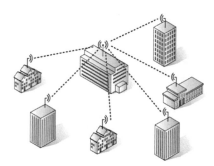

Figure 1.2: Point-to-Multipoint Network

base station back-bone links, etc. A single network is quickly deployed, but neither is this scheme scalable to networks serving a large number of subscribers, nor does it provide for node mobility.

1.1.2   Point-to-Multipoint Networks

In point-to-multipoint networks (figure 1.2) a particular radio, the base station[1] (BS) plays a key role. Here, a wireless link is shared between the BS and multiple client[2] radios at diverse sites. Adding a new subscriber usually means adding a new client radio, while no modifications to the BS or existing clients are necessary. According to the hierarchical network structure, the link from BS to clients is called "down-link", while "up-link" denotes the reverse direction. Usually, direct communication is only possible between the BS and individual subscribers: typically, no two subscribers can communicate directly with each other, even if within radio range.

Radio resources are shared between several subscribers. Employing packet-switching, such networks can be designed to benefit from the statistical multiplexing gain. Peak burst rates can be made available to individual subscribers without having to dedicate the entire bandwidth to them, all the time. Obviously, such kind of a network scales extremely well **on the down-link**. Examples for such networks – serving millions of subscribers – include the Global Positioning System (GPS) [1], terrestrial television broadcasting services, and second/third generation (2G/3G) cellular communications [2–5]. However, clients must be within range of the BS – sometimes even an LOS link is required, for instance in satellite networks. In other words, coverage is limited to a single wireless hop around the BS.

1.1.3   Multipoint-to-Multipoint Mesh Networks

Resembling the structure of the (wired) Internet [6–8], routed mesh networks are probably the most flexible and cost-efficient way of providing interactive broadband services to dense pockets of homes, offices and mobile users. Each radio in the network has equal capabilities and becomes part of the infrastructure, routing data through the wireless mesh network, similar to the wired Internet. This paradigm shift brings **peer-to-peer** communications to the wireless world.

With cellular networks, spectrum has to be licensed, BSs need to be build up on leased ground, the network must be permanently monitored and actively managed by its operator, etc. No such action is required in ad hoc networks. Only if access to public networks, such as the Internet, or

---

[1]Sometimes also called master, access point, central coordinator, etc.
[2]Alternate terms include: subscribers, users, mobile stations, slaves, etc.

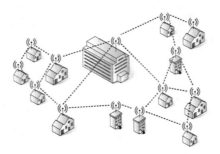

Figure 1.3: Multipoint-to-Multipoint Mesh Network

the Public Switched Telephone Network (PSTN) is desired, BSs are required. However, they are only required at distinct "points-of-presence" – not necessarily arranged in a cellular fashion.

The present work contributes to the development of multipoint-to-multipoint mesh networks. A number of challenges, benefits and disadvantages, are presented in the course of this chapter and, naturally, in the whole remaining work. Therefore, more about this wireless network architecture is deferred to subsequent sections.

1.2   The ad hoc Networking Paradigm

1.2.1   Terminology

The terms mobile ad hoc network (MANET), multihop wireless network, packet radio, wireless mesh network, etc. are often used synonymously to describe a network that does not rely on any preexisting, fixed infrastructure. Typically, stations join or leave the network dynamically (at random) due to mobility, capture and fading effects on the wireless channel, power source failures, etc. This may happen without any prior indication, and perchance without any negative impact on other stations' communication. A thorough introduction and survey of state-of-the-art in ad hoc networking technology is provided in [9], and references therein.

It is worth clarifying that, in this dissertation, the notion "ad hoc" is meant in exactly this way, as there is some confusion in the literature about this term[3]. That is, "ad hoc" shall refer to a self-organizing wireless **multihop** network. In particular, the present work is meant as a contribution to the development of emerging large-scale metropolitan and perhaps even wide area mesh networks. These shall be able to provide integrated broad-band data and interactive audio/video services. Also note that the expressions station, node, and terminal are used interchangeably.

1.2.2   History and Drivers

Military Roots

The ad hoc paradigm has originally been developed for military applications (battlefield communications) in the early 1970s, and has made continuous progress since then. Development in this sector has been pushed by numerous research projects sponsored by the Defense Advanced Research Projects Agency (DARPA), which is the central research and development organization

---

[3]To be more specific, the IEEE 802.11 standard defines ad hoc networks as networks "composed solely of stations within mutual communication range of each other" [10], thereby confining them to a single-hop topology

for the US Department of Defense (DoD). A large number of projects has been supported in the framework of the PRNET [11], SURAN [12] and GloMo [13] programs.

Civilian Drivers

In recent years, ubiquitous computing has become an everyday technology to many of us. Mobile phones, personal digital assistants (PDAs), notebook PCs, smart devices, etc. demand for inter-connectivity. Therefore, the salient feature of ad hoc networking, namely its ability to operate in different and differing environments, has made this technology so attractive for civilian applications. It is a perfect match for situations, where propagation and network operational conditions cannot be predicted during the network design stage.

1.2.3   Advantages of Multihop Networks

Moreover, the multihop paradigm might, very well, be an appropriate answer to the ever growing demand for network capacity – mainly for three reasons. First of all, it solves the data rate vs. distance dilemma: assume constant output power at the antenna (limited by regulatory authorities), and a specified bit error rate (BER); then, it is possible to use higher rate modulation schemes [14] (16-QAM, 64-QAM) for transmissions to nearby receivers, while more robust (and less efficient) schemes (QPSK, BPSK) are required to reach far-away stations.

Second, spectral efficiency is dramatically increased compared to 2G/3G cellular systems, because "subscriber terminals do not 'shout' at a centralized base station, but rather whisper to a nearby terminal that routes the transmission to its destination" [15]. In this perspective, stations cooperate instead of competing for scarce wireless resources. As a matter of fact, reuse of radio resources increases, while overall battery consumption and RF radiation within the collective are significantly reduced.

Last not least, the resulting mesh structure creates the possibility to route around blind spots, and circumvent congested areas in an adaptive fashion. For example, such an approach has recently become a viable option for reach and coverage extension around hotspots [16]. In contrast to cellular systems, additional users enhance rather than stress network capacity – at least to some extent.

1.2.4   Challenges, Problems, and Disadvantages

However, such telecommunication systems combine the troubles stemming from a meshed network structure with the exceptional impairments associated with wireless communications. Hence, this class of "pure wireless" networks constitutes one of the most challenging fields of research and development, today. Besides the fact that **all** communication is wireless, the potential for rapid node movements and the lack of a central coordinator are further origins of inherent problem complexity. These are also the limiting factors that currently restrain the ad hoc paradigm to local, personal, and body area networks with a very limited network diameter[4]. In addition, the end-to-end delay increases steadily with growing hop count. This might not be a problem in streaming applications, but interactive applications are susceptible to latency. For mobile devices, energy considerations play a key role in almost every part of the system design [17].

---

[4]The maximum number of hops between any two nodes, when only the shortest paths are considered

1.2.5 Flat vs. Hierarchical Architectures

Organization of ad hoc networks may be either hierarchical[5] or flat [18]. Hierarchy is created dynamically by partitioning the whole network into groups, so-called **clusters**. Within each cluster, one station is elected as the cluster head. Cluster heads take responsibility for routing, channel assignment, etc. for all cluster members. For example, in hierarchical routing, traffic between two nodes that are in different clusters is always through the cluster heads. This can easily lead to sub-optimal routes.

Flat networks have some advantages over clustered ones. First of all, multiple paths between two stations may exist, such that the traffic can be spread among multiple routes. This effectively reduces congestion and avoids potential bottlenecks. Moreover, there is no "single point of failure", caused by the cluster heads. It is also difficult to maintain the hierarchical structure in presence of node failures, movement, etc.

1.2.6 Revolutionary vs. Evolutionary Architectures

Besides the revolutionary architecture described above, also evolutionary proposals exist, which allow for a smooth transition from 2G/3G cellular to 4G+ meshed networks. These include complementary ad hoc technology for intra-cell communication in existing 2G and 3G cellular networks. One such proposal includes A-GSM, which aims to improve coverage by multihop relay over intermediate handsets [19].

Evolutionary concepts often incorporate a mixture of classical and new approaches. For instance, in a GSM system with ad hoc capabilities within cell boundaries, resource management may be performed in a centralized fashion by a cell's base station [20]. In contrast, revolutionary "pure wireless" solutions require a fully distributed regime, like the novel TBSD-RRM to be introduced in chapter 4.

1.2.7 Applications

Multihop ad hoc networks can be seen as an enabling technology for a vast number of potential applications. In this perspective applications are almost countless; nonetheless, a few are going to be presented next.

Telecommunications

Wireless broadband Internet access for dense residential areas has already been mentioned and the benefits of scalability, efficient spectrum utilization and rapid deployment have been underlined. A large percentage of phone calls is usually placed within a confined area, for example in the same city: in a multihop ad hoc network, such calls could be served without stressing existing extrinsic infrastructure. Disaster recovery is a further good example for the potential use of ad hoc networking.

Networked Neighborhood

Another interesting application would be the "networked neighborhood", which would allow users to share files, play interactive games, etc. with their friends in the neighborhood. Once a "community network" has been established, it can be used for multiple purposes. For example, such a network would be ideal to monitor intelligent surveillance equipment, share videos, etc.

---

[5]Not to confuse with inherently hierarchical wireless networks, such as 2G/3G cellular networks

Intelligent Transportation

Vehicular ad hoc networks are also a hot topic of current research. Car-mounted transceivers could be used for vehicle-to-vehicle communications. Currently, an international consortium consisting of companies, regulatory authorities and standardization bodies is developing recommendations for a so-called Dedicated Short Range Communications (DSRC) system. The DSRC system shall employ multihop ad hoc technology for vehicle-to-vehicle, vehicle-to-roadside and roadside-to-world communications.

Until now, more than thousand applications have been proposed, and the list is still growing. Just to mention a few: approaching emergency vehicle warning; emergency vehicle video relay; imminent collision warning; cooperative adaptive cruise control; traffic management (infrastructure based); stop light assistant (infrastructure based); green light, optimal speed advisory; toll collection; traffic information; enhanced route planning and guidance; access control; parking lot payment; fleet management; access control; and many others. . .

Sensor Networks

From an application point of view, sensor networks can be interpreted as distributed data bases. A potentially large number of tiny sensor nodes, featured with RF modules, could be deployed on (possibly hostile) terrain. These could be subsequently used to collect information about environmental conditions (temperature, humidity, pressure, etc.), detect movement, and so on. More information on sensor networks is provided in chapter 5.

1.3   Quality-of-Service

Quality-of-service (QoS) is a term that can generally be seen from two perspectives: the network's viewpoint, and the user's viewpoint. From a network perspective, QoS denotes the ability to provide traffic differentiation, such that different users running different applications can be offered different levels of service, according to application requirements. To the user, QoS represents the network's ability to maintain certain guarantees on performance, delivery, and reliability [21].

Traditionally, only fixed wired networks with large bandwidths and low error ratios had been addressed. Latterly, this apparatus has been enhanced to also cover cellular systems embracing a final wireless hop, from BS to subscriber, like with Wireless ATM [22]. Recently, such concepts are also applied to the design of ad hoc networks [23, 24], where two challenging aspects make QoS support very difficult: unpredictability of the error-prone wireless links, and frequent changes in network topology.

Both aspects play against the classic QoS principle, where a certain QoS level is negotiated during connection establishment, and the network is expected to maintain this agreed-upon level for the lifetime of the connection. In other words, the network must be able to deliver data, end to end, in a manner that meets the previously contracted QoS level. To some extent, link quality variation can be compensated for by forward error correction (FEC) and automatic repeat requests (ARQ), but changes in network topology are usually difficult to handle.

1.3.1   QoS Parameters

Quality is something that users experience and assess based on individual, subjective measures. Technically, a QoS level may be quantified in terms of:

Application	Symmetry	Data Rate[a]	Delay (one-way)	Jitter[b]	Error Ratio
Voice (conversational)	Two-way	4 – 64 kbps	< 150 ms preferred[c] < 400 ms limit[c]	< 1 ms	< 3% PLR
Voice (messaging)	Primarily one-way	4 – 32 kbps	< 1 s for playback < 2 s for record	< 1 ms	< 3% PLR
Streaming audio	Primarily one-way	16 – 128 kbps	< 10 s	$\ll$ 1 ms	< 1% PLR
Videophone	Two-way	16 – 384 kbps	< 150 ms preferred < 400 ms limit		< 1% PLR
Web-browsing	Primarily one-way	ca. 10 KB	< 2 s preferred < 4 s acceptable		Zero
Bulk data transfer	Primarily one-way	10 KB – 10 MB	< 15 s preferred < 60 s acceptable		Zero
Transactions – high priority	Two-way	< 10 KB	< 2 s preferred < 4 s acceptable		Zero
Transactions – low priority	Primarily one-way	< 10 KB	< 30 s		Zero
Command/ control	Two-way	ca. 1 KB	< 250 ms		Zero
Interactive games	Two-way	< 1 KB	< 200 ms		Zero
Telnet	Two-way	< 1 KB	< 200 ms		Zero
Fax – "real-time"	Primarily one-way	ca. 10 KB	< 30 s/page		$< 10^{-6}$ BER

Table 1.1: Performance Targets for Some Applications According to ITU-T G.1010

[a]For data services, typical amounts of data instead of data rates are specified
[b]A de-jitterizing buffer may be used at the receiver to achieve jitter requirements
[c]Assumes adequate control of talker echo

- **Delay**, i.e. the time that elapses from the instant a packet, cell, or bit is transmitted at the source, until it is received at the destination;

- **Jitter** (or delay variation), which is the variation in the interarrival times of packets, cells or bits belonging to the same data stream;

- **Data rate** (or throughput), which is the average rate at which packets are received at the destination, even under heavy network load;

- **Error ratio**, which is the maximum rate at which corrupt packets, cells or bits may arrive at the receiver.

While these are the most important, and most often used metrics, other metrics are also possible. For example, another important metric (which is usually not negotiable), is **availability**. This is the portion of time the network is operational and responsive to user requests.

Based on these metrics, the International Telecommunication Union (ITU) has specified QoS categories for multimedia services [25] from the user's point of view. This ITU-T recommendation defines performance targets for audio, video and data applications – a few of which are summarized in table 1.1.

1.3.2   Traffic Classes

One of the most comprehensive service architectures has been developed by the ATM Forum, allowing subscribers to specify traffic and performance parameters for their connections [26]. User traffic is differentiated and categorized into classes, based on traffic pattern, QoS requirements and potential use of common control mechanisms. As a result, each class is suitable for a certain type of resource allocation. Following categories are available in ATM (simplified):

Constant Bit Rate (CBR)   Real-time traffic with fixed bandwidth, and tight bounds on delay and jitter. Fixed means that the data rate is constant on a **per connection** basis, that is, different connections may have different data rates. Applications: voice, fixed bit rate coded video, circuit emulation.

Variable Bit Rate (VBR)   Intended for bursty sources with varying bandwidth requirements. While the traffic pattern is defined by this classification, QoS requirements of distinct sources may be different. Hence, two flavors of the VBR service exist:

1. **Real-time VBR**, imposing (possibly tight) constraints on delay and jitter. This type of traffic can usually be statistically multiplexed with other traffic. Applications: variable bit rate coded video, e.g. MPEG4.

2. **Non-real-time VBR**, which does not impose such constraints on delay and jitter.

Available Bit Rate (ABR)   Supports sources that can adjust their information rate to network requirements, within a certain throughput corridor. For example, rate adaptive source coding techniques exist for video compression. Using hierarchical coding techniques, important packets can be distinguished from less important ones. This way, it is viable to always send important packets, while packets carrying detail information are only sent, when bandwidth is available. This results in a graceful degradation of image quality, when the network becomes more and more loaded. Otherwise, there might be users watching high-quality, high-resolution video streams, while others would be completely denied service.

Unspecified Bit Rate (UBR)   Implements a best-effort service with no predefined requirements on delay, jitter, throughput, etc. Bandwidth not occupied by other services is used to transport UBR traffic. When the network is not heavily loaded, UBR traffic may be transferred at peak data rates and low latencies. But when the network becomes congested, no bounds on data rates and delay exist.

1.3.3   Basic Elements of QoS

The QoS subsystem of a QoS enabled network is responsible for controlling access to network resources. For example, during network congestion, a QoS aware system is expected to ascertain that users still receive service at the agreed-upon level. A set of basic QoS mechanisms and techniques has evolved for controlling access to shared network resources, so as to achieve predictable performance[6]. Basic elements of QoS management include:

---

[6]Notice that performance is predictable only in stationary wireless networks

- **Admission Control**. This part determines whether a connection request should be accepted or rejected. The request is permitted, when sufficient network resources are available to carry the requested traffic pattern and QoS profile; otherwise, it is denied.

- **Congestion Avoidance**. Anticipates congestion at a node and discards packets when the queue size exceeds some predefined threshold. Usually, packets are dropped selectively, i.e. higher-priority traffic is more likely to be delivered than lower-priority traffic.

- **Traffic Classification**. Assigns different classes to flows with different QoS requirements, thereby allowing different packets to be treated differently, based on QoS requirements.

- **Traffic Shaping**. This technique is used to reduce burstiness at node input, in order to ensure that the traffic does not violate a specified profile.

- **Packet Marking**. Assigns arriving traffic to a QoS class.

- **Packet Scheduling**. Uses a queuing scheme to schedule packet transmissions. Primary schemes include strict-priority queuing, class-based queuing, and weighted fair queuing.

1.3.4   QoS Implementations

Both, reservation-oriented and reservation-less QoS implementations exist. In reservation-oriented QoS, resources are explicitly reserved for a given data stream. ATM and the Resource Reservation Protocol (RSVP) [27] are examples for such an implementation. Conversely, reservation-less schemes do not explicitly reserve resources. Instead, traffic prioritization is used to let all members of a traffic class receive service according to their priority level.

In this section, a lot of QoS terms have been introduced based on the ATM Forum's terms and definitions. This is because ATM networks **inherently** support QoS. Originally, the Internet Protocol (IP) did not support QoS mechanisms. Latterly, appropriate extensions have been developed addressing these deficiencies, an overview of which is given in [28]. However, "Internet QoS still has a long way to go" [29] due to the Internet's large-scale architecture, diversity in routers, hosts, backbone links etc. Thus, research on related techniques is still ongoing.

1.4   Protocol Terminology

A consistent terminology is used throughout this work to describe protocol architectures. As it is based on the definitions provided by the well-known ISO/OSI model, only a brief review is given in this section. In general, it is assumed that readers are familiar with the concepts of services, service primitives, service and protocol data units (SDUs/PDUs)[7], etc. Anyhow, figure 1.4 shall serve as a reminder for a few typical relationships between service primitives, which are classified into four categories:

- **Requests** are originated at a service user and issued to a lower-layer service provider. For instance, when the transport layer instructs the network layer to deliver an SDU (e.g. a UDP datagram), we call this a request. Similarly, if the MAC instructs the PHY to tune its transceiver to a specific frequency, this is also a request.

---

[7]Recall that SDUs are exchanged with peer layers, while PDUs are passed to adjacent layers

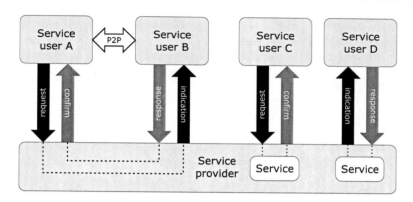

Figure 1.4: Service Primitives and Some Potential Relations

- The service provider may **confirm** the request, i.e. it may report back the result of a request, such as a success notification or failure condition. For a certain kind of request, the choice whether to confirm or not is not optional, i.e. either all requests of the same kind are confirmed or none. Also, a request must precede any confirmation, that is, a confirmation is only issued in reaction to a request.

- By means of an **indication**, a service provider is able to alert a higher-layer service user of some event. For example, if the Internet Protocol has received a datagram, it indicates this occurrence to the UDP layer, passing the SDU along with the indication.

- The higher-layer protocol may in turn **respond** to an indication, e.g. TCP may acknowledge the reception of a fragment, which may then be forwarded to the peer TCP entity in form of a confirmation.

With the help of these service primitives, a peer-to-peer (P2P) connection between protocols on the same layer may be realized (e.g. two UDP instances on remote stations). However, existence of a peer entity is not a requirement, meaning that the same architectural concept may be used to request services in a lower layer, or indicate something to a higher layer service. When we consider IEEE 802.11 as a final example, the PHY reports the result of its clear channel assessment (CCA) measurements in form of `PHY-CCA.indications`. Analogously, the MAC invokes transmission of an MPDU by issuing a `PHY-TXSTART.request`. Again, it should be stressed that all protocols presented in this work – together with their respective implementations – adhere to this unified model.

1.5   Random Access Protocols

A number of random access protocols for sharing wireless channels exist, an overview of the most important ones can be found in [30]. A very basic protocol, pure Aloha, allows all network stations to transmit at will. Obviously, packets can get lost due to co-channel interference when more than one transmission takes place at a time. A station that has transmitted a packet waits for a positive acknowledgment (ACK), and, when no ACK frame happens to arrive, repeats the transmission after a random pause time. As a matter of fact, Aloha is not suitable for heavily loaded networks, but provides fast access to the channel, when network load is rather low. Under heavy load, throughput falls rapidly from its peak at roughly 18.4%, while delay increases steadily.

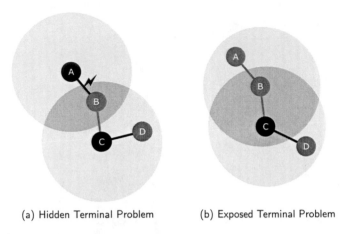

(a) Hidden Terminal Problem   (b) Exposed Terminal Problem

Figure 1.5: Deficiencies of Physical Carrier Sense

1.5.1 Carrier Sense Multiple Access

Carrier sense multiple access (CSMA) is able to utilize the channel more efficiently. Here, the common physical channel is shared by multiple devices in time-multiplex, i.e. stations use the medium sequentially. In contrast to typical TDMA systems, where fixed-length time slots are assigned to individual stations in a centralized fashion, stations contend for transmission opportunities in a CSMA system. Notice that CSMA is also a key element in the IEEE 802.11 MAC protocol to be discussed later on.

Contention is done according to a "listen-before-talk" model: a station with data packets pending for transmission senses the wireless channel for ongoing transmissions of competing stations. If no other transmission is detected, the station starts its own transmission – thereby occupying the shared channel and preventing others from using it. While this is true in theory, practical implications reveal problems.

1.5.2 Near-Far Phenomena in Carrier Sense Multiple Access

Generally, two problems arise in wireless multihop networks employing CSMA [31, 32], known as the hidden and exposed terminal problems. Both stem from the fact that a station willing to start a transmission can sense the medium only at its own location, not at the receiver's site. Both lead to a degradation of network performance. The mentioned problems are illustrated in figure 1.5, and described below:

- **Hidden Terminal Problem.** Station A, willing to transmit an MPDU to station B, senses the medium to be available and starts transmitting. However, receiver B is exposed to the transmission of another station C, which A cannot hear. In this case, the two transmissions interfere and B cannot receive A's transmission.

- **Exposed Terminal Problem.** In the reverse scenario, when B wants to transmit towards A, B is exposed to an on-going transmission from C to D. B could transmit to A without causing interference at station D. Anyhow, B senses the medium to be busy, and thus does not start transmission – and again, throughput is degraded.

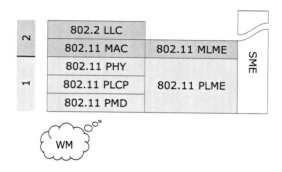

Figure 1.6: Scope of IEEE 802.11

1.6   The IEEE 802.11 Family of Standards

In the remaining parts and chapters, a number of references to the IEEE 802.11 family of standards for wireless LANs are going to be made. The MAC protocol specified in this standard, in particular the so called Distributed Coordination Function (DCF) shall play a key role in the course of this work. The standard, together with its numerous amendments [10, 33–36], covers PHY and MAC layer aspects of wireless LANs. Here, the basic idea is to make the wireless NIC indistinguishable from a wired NIC from the IEEE 802.2 [37] LLC's point of view. In other words, it shall be transparent to the LLC sublayer, whether it operates on top of wired (e.g. according to IEEE 802.3 [38]) or wireless Ethernet. Both sublayers, LLC and MAC, constitute the data link control (DLC) layer. Figure 1.6 gives an overview of the various IEEE 802.11 protocol subsystems and their integration with the IEEE 802.2 LLC. A modular protocol architecture has evolved during standardization of the different physical layer flavors.

1.6.1   IEEE 802.11 Physical Layer Flavors

On the bottom of the stack, the so-called physical medium dependent device (PMD) interfaces to the wireless medium (WM), i.e. the air. In past years, different air standards have evolved, ranging from infrared, over frequency hopping (FHSS) and direct sequence spread spectrum (DSSS) to orthogonal frequency division multiplexing (OFDM). A comparison of the main system parameters for the various RF-based PMD technologies is outlined in table 1.2.

1.6.2   Hardware Addresses

At various occasions, the term "hardware address" is used in this work. At this point, it shall be clarified that said term refers to an universal 48-bit address, according to IEEE 802 [39]. Such addresses are frequently also called MACIDs. The important characteristic about hardware addresses is their global uniqueness: all potential members of a network need to have unique identifiers, in order to coexist in the network.

1.6.3   IEEE 802.11 Medium Access Control

On top of the various PHYs, a single MAC protocol manages access to the wireless channel. The original protocol consists of the DCF, which implements a contention service, and an optional

Standard	802.11 (FHSS)	802.11 (DSSS)	802.11a	802.11b	802.11g	802.11h[a]
Data Rate	1/2 Mbps	1/2 Mbps	6 ... 54 Mbps	1 ... 11 Mbps	1 ... 54 Mbps	6 ... 54 Mbps
Frequency Band	2.4 GHz	2.4 GHz	5 GHz	2.4 GHz	2.4 GHz	5 GHz
Available Spectrum	83.5 MHz	83.5 MHz	300 MHz	83.5 MHz	83.5 MHz	455 MHz
Channel Bandwidth	1 MHz	22 MHz	18 MHz	22 MHz	22 MHz	18 MHz
Independent Channels	3 × 26 sets	3	8	3	3	19
Modulation Scheme	2/4-GFSK 79 channels in set >2.5 hops/s	DBPSK/DQPSK 11-chip PN code (Barker sequence)	BPSK/QPSK/ 16-/64-QAM OFDM 64/52 subcarriers (48 data + 4 pilot)	DBPSK/DQPSK 8-chip CCK or PBCC	combination of 802.11a & 802.11b	same as 802.11a
Forward Error Correction Scheme	-	-	convolutional code 1/2-, 2/3-, 3/4-rate K=7 (64 states)	-	same as 802.11a	same as 802.11a
Output Power, EIRP	100 mW	100 mW	30 mW	100 mW	100 mW	200/1000 mW
Typical Indoor Range	50 m	50 m	10 m	30 m	30 m	40 m
Compatibility	-	-	-	802.11 DSSS	802.11b	802.11a
Year of Approval	1997	1997	1999	1999	2003	2003

Table 1.2: Various IEEE 802.11 PMDs and Their System Parameters (Most of Europe)

[a]Note that IEEE 802.11h also adds dynamic frequency selection (DFS) and transmit power control (TPC) to avoid interference with radar systems

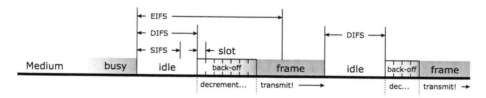

Figure 1.7: Basic Channel Access under DCF Regime

Point Coordination Function (PCF) working on top of the DCF. The PCF provides a contention-free service, but requires an access point taking the role of a central coordinator to perform polling in the contention-free period. Hence, the PCF is out of the scope of this work, as it is not suitable for ad hoc networking. The MAC layer provides three service primitives to the LLC, namely the

- `MA-UNITDATA.request`, which requests a transfer of a MAC service data unit (MSDU) from the local LLC sublayer to one or more peer LLC entities;

- `MA-UNITDATA.indication`, which transfers an MSDU received through the wireless medium to the local LLC sublayer;

- and `MA-UNITDATA-STATUS.indication`, which has only local significance and provides the local LLC entity with status information reporting the results of a preceding `MA-UNITDATA.request`.

Distributed Coordination Function

The DCF employs a sophisticated version of CSMA, namely carrier sense multiple access with collision avoidance (CSMA/CA), enhanced by random back-off timers and immediate positive ACKs. In addition, large frames are optionally partitioned into smaller fragments, which reduces collision probability and the "cost" of fragment loss. Attention has been payed to reduce collision probability between competing stations while designing the CSMA/CA protocol. The highest probability of collision exists at the point, when the medium becomes idle after it had been busy before. The reason is that multiple competing stations could have been waiting for the medium to become idle again. Therefore, a random back-off procedure has been introduced to resolve such conflicts. Otherwise, perpetual collisions would be programmed, since stations would "synchronize" on the busy-to-idle transition. The basic access method under DCF regime is illustrated in figure 1.7.

The deficiencies of physical carrier sense have been mentioned earlier. In order to combat at least the hidden terminal problem, a virtual carrier sense mechanism is also part of the DCF. This mechanism includes the distribution of medium reservations announcing the imminent use of the channel. One way to distribute such information consists of request-to-send (RTS) and clear-to-send (CTS) message exchanges, prior to the actual data transmission. RTS and CTS frames contain a duration field that specifies the period of time necessary to transmit the actual data plus a corresponding ACK frame. The originating station transmits the RTS and waits for the destination to reply with a CTS. Since both frame types contain the duration, all stations receiving either the RTS or the CTS recognize the impeding use of the shared medium, and shall defer their own transmissions until the end of the announced period.

As a further means of distributing medium reservation requests, all directed frames also include the duration field. In this case, the duration will cover the period of time required to receive the mating ACK frame, or – in case of a fragment burst – the time required to receive the ACK, transmit the next fragment and receive its corresponding ACK.

Another benefit of the RTS/CTS mechanism is related to the short length of these frames: a sequence of RTS/CTS can be used to perform fast collision inference. If no matching CTS is detected in response to the RTS, only the short RTS frame needs to be retransmitted, not a lengthy data fragment. However, the RTS/CTS mechanism, as well as immediate ACK frames, cannot be used with broadcast and multicast receiver addresses, because of the potentially large number of stations responding concurrently (well-known ACK-explosion problem).

Furthermore, the RTS/CTS mechanism may not always be justified due to the additional overhead. The mechanism may be particularly inefficient, when data frames are comparably short. Therefore, a threshold allows using RTS/CTS either always, never, or only when frame lengths exceed said threshold.

Interface Space The time span between frames transmitted on the channel is called the interframe space (IFS). For the medium to be considered physically idle, a station has to perform carrier-sense on the medium for a certain IFS. Different priority levels are provided by means of different IFS lengths: the shorter the IFS, the higher the probability for a station to win the contention. Four IFS lengths have been defined[8]:

- **Short interframe space (SIFS)**. The SIFS is used for ACK and CTS response frames, as well as second and subsequent frames of a fragment burst. Since SIFS is the shortest IFS, it shall be used when a station has seized the medium and needs to keep it for completion of a frame exchange sequence. This prevents other stations from interrupting the frame exchange sequence in progress, because they are required to wait for the medium to be idle for a longer time gap.

- **PCF interframe space (PIFS)**. Used when operating under PCF.

- **DCF interframe space (DIFS)**. The DIFS is the regular IFS that all stations must obey to, when operating under DCF to transmit data frames. It is longer than SIFS, and thus does not interrupt high-priority message exchange sequences.

- **Extended interframe space (EIFS)**. The EIFS, which is considerably longer than DIFS, shall be used instead of DIFS, whenever a station recognizes reception of a corrupt frame. As soon as a valid frame is received, the DIFS becomes the default IFS for that station again. If, for example, another station has acknowledged what was – to this station – an incorrectly received frame, this will terminate the EIFS period preemptively, provided that at least the ACK frame is received successfully.

Random Back-off Each station maintains a random back-off counter. Whenever it has to perform the random back-off procedure, it first checks if the back-off counter is zero. If so, it generates a random number counting time slots[9] that are added on top of the usual IFS (DIFS or EIFS). This random number can range from zero to a variable upper bound, called the contention window (CW). Whenever a retransmission is required, the CW is grown exponentially so as to improve stability of the access protocol under high network load.

---

[8]Notice that a QoS extension is under development, which will add further IFSs for real-time services [40]
[9]The duration of a time slot depends on the physical layer in use

Stations use physical carrier-sense to assess activity on the channel. If there is no activity during a back-off slot, the back-off counter is decremented by one. But if the medium is determined to be busy at any time during back-off, then the counter is not decremented; instead, it maintains its count. Afterwards, if the medium has become idle for more than the usual IFS (DIFS or EIFS), the back-off procedure is resumed. Transmission commences, when the counter reaches zero.

On one hand, increasing the CW on consecutive retry attempts may lead to better utilization of the spectrum, because those stations that are likely to complete their transmissions successfully are privileged by lower CWs, and thus spectrum is occupied by those stations that can actually "use" it. On the other hand, this presents a source of inherent unfairness, as it favors stations that have been successful in the past over those which have not.

In general, fairness is enforced by requiring stations that have transmitted an MSDU to perform the back-off procedure, such that frames transmitted from the same originator are always separated by at least one back-off interval. This will provide transmission opportunity to other stations with pending MSDUs. So, a station transmitting a continuous data stream cannot use the medium exclusively thereby causing denial of service to other stations. More detailed information on the DCF is also provided in sections 3.1 and 6.7.1.

1.7   Further Related Technologies and Research Projects

1.7.1   Bluetooth (IEEE 802.15.1)

Bluetooth [41, 42] started off as a cable replacement technology and soon became the de-facto standard for wireless personal area networks (WPANs). Nowadays, there is barely any mid-class mobile phone not equipped with a Bluetooth interface. A wealth of compatible headsets, medical surveillance devices, printers, barcode readers, GPS receivers etc. is available.

The system operates in the license-exempt 2.4 GHz band (like IEEE 802.11), using a frequency hopping transceiver and TDMA/TDD as multiplexing/duplexing scheme. The elementary network structure is called a piconet, which consists of two or more devices occupying the same physical channel. In each piconet exactly one device takes the role of a master, which is responsible for clock distribution (synchronization of slaves). The hopping sequence is derived from the master's burnt-in hardware address and its clock, so as to allow coexistence of multiple co-located piconets. While any Bluetooth node can take the role of the master, it can do so for only one piconet at a time, i.e. it cannot be the master of two or more piconets. However a single device may participate concurrently in two or more piconets, thereby forming a so-called scatternet. It does this as a slave, on a time-division multiplexing basis.

Communication is of ad hoc nature, in the sense that the network is typically created in a spontaneous manner: no formal infrastructure is required, and the network is limited in temporal and spatial extent. However, the number of devices in the network is also very limited, and the entire network is controlled in a master/slave fashion. Thus, concepts developed in this context hardly scale to wide area, large scale multihop wireless networks, which are in the primary focus of this work.

1.7.2   IEEE 802.16: WirelessMAN and WirelessHUMAN

Wireless metropolitan area networks are in the scope of the IEEE 802.16 working group. The basic **WirelessMAN** standard [43] specifies an air interface for a fixed point-to-multipoint broadband wireless access system, consisting of appropriate PHY and MACs. The system is able to simultaneously support multiple differentiated services (with respect to QoS requirements). The MAC layer is capable of supporting a number of PHY layer implementations optimized for the frequency bands of the application. The standard also includes a particular PHY broadly applicable in the range 10 – 66 GHz, which is intended for LOS links in licensed frequency bands only. Modulation schemes and FEC rates are adaptive on a burst-by-burst basis, enabling to trade-off capacity vs. robustness in real time. Multiple access is organized in TDMA fashion, while the duplex scheme may be either FDD (static asymmetry) or TDD (dynamic asymmetry).

The resulting network architecture resembles that of 2G/3G cellular wide area networks: immobile base stations are connected to public networks, offering their services to subscriber stations, which are also stationary. Therefore, this standard is suitable as a "last-mile" access, providing public network access to residential and office buildings. However, subscriber stations are not able to communicate directly with each other, despite being within radio range.

Amendment project 802.16a [44] added an additional physical layer for 2 – 11 GHz non-line-of-sight (NLOS) links, also to be used in licensed frequency bands only. Besides the previously defined single carrier (SC) modulation, two additional physical layer flavors have been specified: the first one adds OFDM support, while the second one enhances the SC modulation so as to combat NLOS multipath effects. In addition, this amendment provides the ability to operate a wireless MAN in license-exempt bands employing dynamic frequency selection (DFS). This particular flavor is entitled Wireless High-Speed Unlicensed Metropolitan Area Network, or **WirelessHUMAN** in short. Moreover, an optional mesh topology is defined, which allows direct subscriber-to-subscriber communications without needing dedicated base stations. That is, base stations are only required to provide backhaul links.

Also new in 802.16a, two scheduling schemes are available, which can be used for bandwidth allocation: a distributed, and a centralized scheme. In the so-called "coordinated distributed scheduling mode", two-hop neighborhood information is used to schedule transmissions in a manner that they do not collide with each other. Also, the distributed scheme does not rely on the existence of a BS. In these perspectives, the distributed scheme shows some similarity with the TBSD-RRM protocol presented in chapter 4. However, TBSD-RRM is by far more flexible.

1.7.3   IEEE 802.20: Mobile Broadband Wireless Access

While IEEE 802.16 provides a solution for fixed broadband wireless access, an emerging working group, IEEE 802.20, is concerned with support for mobile devices [45]. The air interface will be optimized for IPv4 and IPv6 traffic. It should be noted that, at the time of writing, this project had been in a very early stage. It presents a work in progress, and hence, is subject to change. Nonetheless, basic (preliminary) architectural parameters are presented in table 1.3, as certain ideas presented throughout this work are also applicable to such a kind of system.

Systems based on IEEE 802.20 are intended to provide ubiquitous mobile broadband wireless access based on a cellular network architecture, consisting of macro/micro/pico cells. These systems shall support NLOS outdoor and indoor scenarios. Provisions are made, to provide link-level QoS mechanisms, based on Internet QoS standards. In addition, the link-level QoS structure shall provide sufficient capabilities to conform to end-to-end QoS architectures, for example those mentioned in section 1.3.4.

1.7.4   Routing Protocol Standardization: IETF/MANET

The Internet Engineering Task Force (IETF), the principal protocol standards development body for the Internet, has launched a working group for mobile ad hoc networks, called MANET. The

Characteristic	Target Value
Mobility	Vehicular mobility classes up to 250 km/h
Sustained spectral efficiency	> 1 b/s/Hz/cell
Peak user data rate (downlink)	> 1 Mbps[a]
Peak user data rate (uplink)	> 300 kbps[a]
Peak aggregate data rate per cell (downlink)	> 4 Mbps[a]
Peak aggregate data rate per cell (uplink)	> 800 kbps[a]
Air link MAC frame round-trip time	< 10 ms
Bandwidth	e.g. 1.25 MHz, 5 MHz
Maximum operating frequency	< 3.5 GHz
Duplexing scheme	Supports FDD and TDD
Spectrum allocations	Licensed spectrum
Security support	Advanced Encryption Standard (AES)

Table 1.3: Key IEEE 802.20 Parameters

[a]Targets for 1.25 MHz channel bandwidth, i.e. 2×1.25 MHz FDD or 2.5 MHz TDD

main purpose of this group is to standardize IP routing protocol functionality suitable for wireless routing applications. For example, AODV [46] is one such approach that has recently reached the "experimental protocol" stage. Both, static and dynamic topologies are addressed. As part of this effort, the group will also address aspects of security and congestion control in the designed routing protocols.

1.7.5   CMU Monarch

In the Monarch project [47] at Carnegie Mellon University (CMU), networking protocols have been developed, which allow seamless wireless and mobile host networking. The scope of CMU Monarch research included protocol design, implementation, performance evaluation, and usage-based validation, spanning areas ranging from L2 to L6. A lot of work has been done in the area of routing support for mobile hosts operating in a large internetwork, including the design of Mobile IP [48].

A wireless testbed has been developed as part of the Monarch project to enable the evaluation of ad hoc networking performance in the field [49]. The actual ad hoc network was comprised of five moving car-mounted nodes, and two stationary nodes. At that time, IEEE 802.11 compliant cards were not available, therefore AT&T WaveLAN-I radios have been used. These were operating at 900 MHz and provided 2 Mbps through DSSS modulation. Furthermore, wireless extensions to the ns network simulator [50] have been introduced by CMU Monarch. This simulator is often used by researchers to investigate multihop ad hoc networks.

1.7.6   MIT Roofnet

The Massachusetts Institute of Technology (MIT) has created an experimental multihop mesh network based on IEEE 802.11b compliant hardware [51]. This testbed is known as the MIT Roofnet. It consists of about 50 nodes installed in apartments in Cambridge, Massachusetts. Each node is in radio range of a subset of the other nodes, and multihop forwarding is used to reach the rest of the nodes. A few gateway nodes provide access to the wired Internet.

Roofnet is self-organizing, in that it requires no configuration or planning. There is no need to allocate IP addresses, direct the rooftop-mount antenna, etc. At the time of writing, the longest routes were four hops long. Latencies typically accumulated to dozens of milliseconds – even for long routes; typical throughputs were in the range of several hundred kbps up to a few Mbps. From the algorithmic point of view, primary focus is on efficient routing in multihop networks. A property of unplanned networks is that each node can route packets through any of a large number of neighbors. However, radio link quality varies largely from neighbor to neighbor, and time to time; finding the best multihop routes through a rich mesh of variable quality links turned out to be a challenge [52].

1.7.7 Nokia Rooftop

The idea of mounting wireless routers on top of subscribers' roofs was not new. In late 1999 Nokia Corp. acquired Rooftop Communications, Inc. who had followed a similar approach. However, they did not use 802.11 hardware, but developed an embedded system based on a FHSS radio in the 2.4 GHz spectrum employing 2-/4-GFSK – thereby providing up to 2 Mbps of data rate. A proprietary "Nokia AIR Operating System" had been designed, comprising channel access protocols, reliable link and neighbor management protocols, wireless multihop routing and multicast protocols, as well as standard Internet protocols for seamless integration with the (wired) Internet. While this approach gained considerable attention [53], Nokia discontinued its wireless router activities in 2002, probably due to technical problems.

1.7.8 MeshNetworks

A test network has been set up by broadband wireless start-up MeshNetworks, Inc. providing data rates up to a few Mbps at vehicular mobility. Even at (US) "speed-limit speeds" [54], the network was able to sustain download data rates of 500 kbps. MeshNetworks offers software that augments existing MAC and LLC layers with meshing control components for automatic route discovery, ad hoc route selection, neighbor/link metrics, etc.

In addition, a proprietary air interface has specifically been designed for wide area multihop ad hoc networks. It provides multiple channels: three data channels and a control channel in the 2.4 GHz band. Key to MeshNetworks' Quadrature Division Multiple Access (QDMA) technology is a patent [55] that describes a communications method for a CDMA system without a base station. Moreover, QDMA radios provide built-in positioning methods that do not rely on GPS. These can be used to offer location-based services for military, public safety, intelligent transportation, m-commerce applications, etc.

1.7.9 WINGS

As part of the DARPA GloMo program, the Wireless Internet Gateways (WINGS) [56] project extended the Internet to multihop wireless networks. In cooperation with Rooftop Communications, Inc. wireless gateways were designed, analyzed, implemented, and tested, which were intended to enable the seamless integration of self-organizing wireless networks with the Internet. As a consequence, a lot of developments have been reused in Nokia's wireless routers. WINGS is based on a two-tier approach. Long-range radios, which are transportable and reside in vehicles, tents, or on roof tops, are used to establish dynamic backbones. Short-range radios, which are low power and can be hand held, serve as access points for mobile users. Mainly, a number of networking issues were addressed. For example, Dijkstra's shortest-path algorithm [57] has been used over a hierarchical graph of the network to determine routes for packet forwarding.

1.7.10   MMWN/DAWN and UDAAN

The Density and Asymmetry-adaptive Wireless Network (DAWN) was a research project having been conducted at BBN Technologies as part of the DARPA GloMo program. In mobile wireless networks, variations in density and asymmetry pose several problems in network connectivity, scalability, real-time multimedia transport, etc. The DAWN project aimed at developing a modular set of density- and asymmetry-adaptive mechanisms that were to provide flexible, survivable, and scalable solutions to such problems. The DAWN system built upon and complemented BBN's existing Multimedia Support for Mobile Wireless Networks (MMWN) [23], and extended its ability to operate under a wide range of network densities and link asymmetries.

Another suite of mechanisms, entitled Utilizing Directional Antennas for ad hoc Networks (UDAAN) has also been developed at BBN. This work was among the first combining ad hoc networking with directional antennas. Major UDAAN innovations included directional medium access control, neighbor discovery using directional antennas, and topology control using antenna pointing. From these innovations, significant payoffs have been achieved [58]: Increased network capacity, decreased packet latency, increased connectivity, and greater multicast efficiency.

1.7.11   Terminodes

A long-term (2000 – 2010) research project has been defined in Switzerland, aiming at studying and prototyping large-scale self-organized mobile ad hoc networks [59, 60]. This group has coined the expression "terminode", describing the fact that user equipment features (terminal) are integrated with wireless router capabilities (network node). The goal is to study fundamental and applied questions raised by new generation mobile communication and information services, based on self-organization.

In contrast to many other projects, a broad set of researchers (about 30 faculty members and 70 PhD students) are brought together, in order to study most aspects of self-organizing, distributed communication and information services. These investigations range from fundamental mathematical issues (statistical physics based analysis, information and communication theory) to networking, signal processing, security, distributed systems, software architectures and economics. In other words, Terminodes research spans from L1 to L7 and consequently a huge number of publications are available addressing various issues (e.g. geodesic routing [61], GPS-free positioning [62], and many more). Thorough information can be found on the Internet, at http://www.terminodes.org.

1.7.12   4G Research

Besides infrastructural improvements, most research forums driving 4G research – including the Fourth Generation Mobile Forum (4GMF), the Wireless World Research Forum (WWRF), and the Mobile IT Forum (mITF) – also provision multihopping as an essential feature of future 4G systems [15, 63, 64]. Further aspects of 4G research include convergence of 3G cellular and WLAN/WPAN technologies, adaptive modulation and coding, multiple-input multiple-output (MIMO) space-time coding, beam-forming antennas, and open software defined radio (SDR) architectures – among others. Generally, ad hoc technology should be seen as a complement to existing heavily infrastructure-based networks, i.e. cellular, satellite and terrestrial broadcasting networks [65, 66]. Reconfigurable radios will allow users to join a diverse set of networks, ranging from multihop WPANs, hotspot WLANs, over cellular networks, to ad hoc networks with proprietary air interfaces.

## CHAPTER 2

Design of A New Architecture for Wireless Multihop Networks

As we have learned from the previous chapter, there are a lot of choices to take, when designing a multihop ad hoc network. In the course of this chapter, the scene will be set for the remaining parts of this work. Starting with an analysis of shortcomings of current state-of-the-art, the novel idea of L2 cut-through switching in wireless multihop networks is proposed. The motivation behind this architecture stems from the desire to let the network provide QoS **by design**. In contrast to many other proposals, emphasis is on a technically feasible system.

2.1   Current Best-Practice

Many of the technologies and protocols used in practical multihop ad-hoc network implementations of today are not really suitable fur this purpose. If we just have a look at the prominent IEEE 802.11 MAC, which is widely used in testbeds for multihop networks, this problem becomes evident. Investigations have identified major shortcomings of the 802.11 MAC protocol that prevent it from functioning well in multihop networks [67]. Problems that have been identified include:

- In spite of the standard having paid much attention to the hidden node problem, it still exists in multihop networks.

- The exposed node problem is not tackled at all.

- Carrier sense is usually implemented in such a way that the sensing/interfering range is typically larger than the communication range. The larger interfering range worsens the hidden node problem, while the larger sensing range intensifies the exposed node problem.

- The back-off scheme always favors successful nodes, which causes unfairness even in a single-hop scenario.

Improvements of the legacy 802.11 are on their way, providing QoS support in wireless LANs [40]. While these will be helpful to address some of the problems, they cannot eliminate all of them, because they merely focus on a single-hop environment.

   In the following subsection, the reasons for this situation will be identified based on a detailed study of the actions involved in the data transmission and forwarding process. Focus shall be

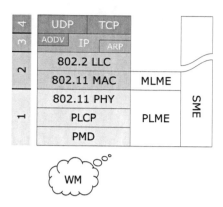

Figure 2.1: Protocol Stack for a Typical MANET

on sources of delay, which prevent current systems from providing services with low end-to-end latency requirements across several hops, such as voice telephony. The complete stack discussed in the following subsection has been implemented into the simulation environment to be presented in part II, and has been analyzed in detail.

2.1.1   Seven Layers of Delay

Let us consider typical design choices for a practical implementation of a mobile ad hoc network (MANET). Figure 2.1 illustrates an ISO/OSI stack representation of the protocols usually involved in communications. At the top of the stack, i.e. at layer seven (L7), an application (which is not included in the figure) selects either the User Datagram Protocol (UDP) [68] or the Transmission Control Protocol (TCP) [69] as an L4 transport protocol. This selection depends on the application's requirements regarding transport reliability, multi-cast capability, and so forth. At L4, the data provided by the source application is split into IP packets and forwarded to L3.

Layer Three - Network

The IP implementation at L3 receives a packet request and puts the associated data into its queue of pending packets. In order to deliver the packet, IP needs a route to the destination station. At this point, we assume the routing protocol to be AODV [46], one of the best-performing protocols for MANETs. A new C++ implementation has been added to the simulation and demonstration framework in support of system-level simulation capabilities. AODV maintains a route table entry for each destination IP address, containing the IP address of the next hop towards the desired destination, the distance (in hops), the route's lifetime (mobility support), and other information.

If a route table entry does not exist for the desired destination, AODV invokes its route discovery process, which is a quite expensive operation in terms of delay. As part of the route discovery, AODV floods the network with route request messages, according to an expanding ring search. This is done by sending UDP datagrams as limited broadcasts (which have a predefined hardware address mapping). Notice that this involves all the steps and delays associated with lower layers, which are going to be mentioned below.

Eventually, a route becomes available in form of a route reply message, meaning that – for a given destination IP address – the station learns the next hop's IP address towards that destination. However, for a directed transmission, the node also needs the next hop's globally unique hardware address. Consequently, an address resolution must be performed, which again incorporates message exchanges. The address resolution protocol (ARP) [70, 71] broadcasts an ARP request using a DLC frame that contains the desired logical address. The addressee receives this frame, recognizes its own logical address and responds with an ARP reply bearing the required address mapping.

Another factor, which may lead to increased latency while waiting for a route and hardware address mapping, is related to the messages being sent as broadcasts. Broadcasts are not protected by L2 ACK frames, and the issuer of a broadcast will never know how many receivers have been reached by the transmission. Depending on the distance from source to destination, route discovery may require up to several seconds.

Layer Two - Data Link Control

As soon as the next hop's hardware address is known, IP forwards the packet down to the L2 DLC protocol. Large IP packets are fragmented into several smaller DLC units. At the upper edge of the DLC layer, the LLC sublayer accepts DLC frames coming from the network. Assume a connectionless mode as defined by IEEE 802.2 – either acknowledged or unacknowledged. In this case, the LLC becomes more or less a simple pass-through layer, adding just a few header bits to the packet. For our analysis, let us consider the associated delay negligible.

At the lower edge of L2, the 802.11 medium access control operates in its ad hoc mode and forms a so called independent basic service set (IBSS). In this mode, the DCF[1] employs CSMA/CA, thereby managing access to a single shared channel, which is cooperatively used by all stations for data transmission.

The delays caused by the MAC protocol are due to the contention-based approach inherent to CSMA/CA: stations with pending MAC frames have to contend for the medium. In a first phase, they request the PHY to listen for on-going data transmissions on the medium, which in turn will report the results of this clear channel assessment (CCA) back to the MAC. Packet transmission starts immediately if the medium is sensed to be idle for a certain amount of time (a few microseconds). In case of a unicast frame, the recipient replies with an immediate positive ACK. This will allow the sender to repeat transmission if no such ACK has been received in time.

If the medium is busy, a station with outstanding packets backs off, i.e. it defers its transmission attempt by a random amount of time. The random approach is required to avoid that stations waiting for a transmission opportunity "synchronize" their attempts to the end of the "medium-busy" condition, which would lead to perpetual packet collisions. This amount, varies between a few tens of microseconds up to a few tens of milliseconds, again depending on the PMD used and the environmental conditions the station is exposed to. The more attempts a station requires for a single packet delivery, the more amount of random time it has to spend waiting. This is done to improve overall network throughput: stations that are exposed to hostile wireless conditions shall not occupy too much of the scarce bandwidth with transmission attempts that are likely to fail in the end. However, this is also a source of unfairness, even in a single-hop environment.

In addition to fragmentation performed by IP, large unicast DLC frames may be further fragmented into smaller portions, which are transmitted as a fragment burst protected by immediate positive ACKs. This is advantageous if the packet size exceeds some threshold. In that case, only fragments that happen to be lost need retransmission and probability of successful delivery is increased, since short frames are less likely to collide than longer ones. Moreover, large DLC frames may be protected by **a priori** medium reservation using virtual carrier sense. This happens

---

[1]Refer to section 1.6.3 for details on the DCF

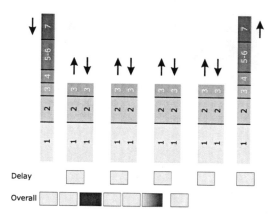

Figure 2.2: L3 Switching in Typical MANETs

via an RTS/CTS sequence, or a "CTS-to-self" – which again requires time. Virtual carrier sense is especially beneficial to combat the hidden terminal problem.

Layer One - Physical

At the bottom of the stack, assume an 802.11 wireless NIC consisting of a physical medium dependent device (PMD), a physical layer convergence protocol (PLCP), and a physical layer interface to the MAC (PHY). These three items make up L1. Depending on the physical layer flavor actually used, i.e. 802.11a/b/g, etc. varying delays through the physical layer may be observed. Transmission delays through the physical layer are caused by interleaving, convolutional coding, data whitening, synchronization preambles, etc. All in all, these delays sum up to a few hundred microseconds and are difficult to circumvent.

2.1.2   Classic Switching and Forwarding

In the previous considerations, we have studied the typical steps required just to transmit a single packet. In a multihop scenario with "store-and-forward" switching in the IP layer, a packet must traverse each intermediate station's protocol from the physical layer, up to the network layer, and back down to the physical layer, and so forth. These steps need to be performed repeatedly at every intermediate hop, as outlined in figure 2.2. Obviously, this incorporates a lot of unnecessary overhead (and thus delay), since it would only be necessary to update some address fields in the MAC frame header, while the payload would never have to be modified.

2.2   An Improved Architecture

In the remaining part of this chapter an improved architecture for a wireless multihop network employing a high-performance switching technique is presented for the first time. This improved approach shares some aspects with related work mentioned below, but also differs in many ways.

2.2.1   Related Work

In the past, a technique known as symbol stream switching has been patented [72], which intended to employ circuit switching of multiple symbol streams arriving in an intermediate station using a switching matrix. A related patent describes the so-called symbol vector switching [73], which is based on similar ideas. However, to the author's best knowledge, both concepts have never been implemented. As a matter of fact, technical feasibility has never been proven.

More recently, a combination of an enhanced 802.11 MAC in conjunction with multiprotocol label switching (MPLS) [74] has been proposed to improve performance of IP-based multihop IEEE 802.11 networks [75]. This effectively shifts switching from L3 to L2 and thus reduces the associated overhead. However, it does not satisfactorily solve the problem that stations have to contend for the shared wireless channel before they can start forwarding pending packets. For 802.11, this means that forwarded packets have to suffer from unpredictable and load-dependent delays due to 802.11's random back-off collision avoidance strategy. In addition to these more analytical considerations, quantitative simulations have also shown that the approach proposed in [75] is only of limited use in a meshed multihop environment [76].

It has been proposed to use multiple NICs in a single station to overcome this problem [77]. One NIC is devoted to provide a channel for signalling purposes, while the remaining NICs offer dedicated channels for user data. This may be a reasonable working proposal for initial research, but it seems to be quite impractical due to the tremendous associated hardware cost. Even though multi-channel chips based on wide-band digital signal processing are also becoming available, packets would still have to be passed from one interface to the other. This incorporates the station's main processor into the data flow, and hence leads to additional delay.

2.2.2   A Novel Architecture Supporting QoS

First of all, there is no self-imposed restriction to solutions that are easily implemented with existing 802.11 NICs. This decision has been taken, because a system with multiple dedicated physical channels available at each station is going to be proposed. Employing existing NICs would in fact require several of them. Instead, the new architecture is supported by two stands: first, the thoroughly studied IEEE 802.11 DCF is used as a random access protocol for the management of a common control channel; second, this channel is used to allocate dedicated channels for data flows with QoS requirements via suitable distributed DCA techniques.

Design and development of a powerful, yet cost-efficient hardware prototype has also been started in parallel to a complete simulation environment, so as to keep in touch with real-world requirements and to come up with practical proposals. Simulator and demonstrator together yield an integrated platform [78] which may be used to verify and optimize new approaches in this emerging sector of research. By the end of this section, it should have become evident that this admittedly high effort pays off, when the advantages and potentials of the presented solution are taken into account.

2.2.3   Protocol Stack

To create an understanding of the overall network architecture it is helpful to give a glance at a station's conceptual protocol stack, which is outlined in figure 2.3. It is worth mentioning that most of this stack has already been implemented into the simulation environment. The simulator now includes models for the wireless medium, the physical layer, IEEE 802.11 MAC, IPv4, UDP, and AODV. Advanced radio resource management, neighborhood discovery, CBR application, and a few other protocols have also been implemented.

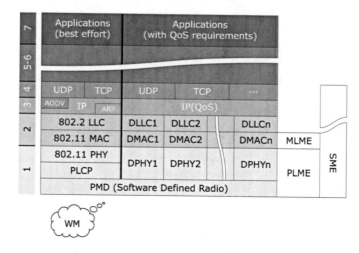

Figure 2.3: Protocol Stack for an Improved MANET

Starting from the bottom layer, the PMD is in our case a software defined radio (SDR). This radio interfaces to the wireless medium and offers more than just one physical channel, either by FDMA, CDMA or combinations thereof[2]. Above the SDR, the protocol stack is divided into two distinct parts. The left side implements a common (shared) random access channel (RACH), while the right-hand stack offers dedicated channels to a subset of neighboring stations. The subset includes those stations that are actually involved in active communications and have established connections beforehand.

2.2.4   Random Access Channel

The PLCP hides the SDR interface's specifics from the generic IEEE 802.11 MAC layer implementation. This physical layer maps to one of the channels provided by the SDR, and tunes the radio to a certain frequency or selects a specific scrambling code common to all stations.

On top of that, the DCF is used to manage access to the shared wireless channel. The DCF essentially realizes a carrier sense ("listen-before-talk") multiple access scheme with random back-off and special control frames (RTS, CTS, ACK). In addition to the DCF, the IEEE 802.11 specification also includes a MAC layer management entity (MLME), which is responsible for establishment of an IBSS, timer synchronization, etc. In the context of this work, special functions have been added to the MLME[3], providing radio resource management for dedicated links and neighbor discovery.

The Internet Protocol has been chosen as the network layer protocol. It makes use of ARP to map logical IP addresses to hardware addresses. Furthermore, AODV has been implemented as a routing protocol. Currently, TCP has not been implemented yet – but UDP has, as AODV makes use of a UDP port to exchange messages with peer instances. The RACH is used to exchange signalling messages that are required to allocate radio resources, explore a station's neighborhood, determine routes, synchronize timers, locate stations, etc. But it may also be used to transmit best-effort traffic in order to improve spectral efficiency. Such kind of traffic would be given less priority than control messages, simply by using an IFS longer than the usual

---

[2]In addition, TDMA may be implemented in the MAC layer
[3]These management functions are detailed in chapter 4

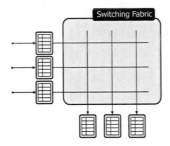

Figure 2.4: Switching Fabric (Cut-through Switching)

DIFS for such best-effort frames – or, vice versa, by using a PIFS for high-priority management frames.

2.2.5 Dedicated Channels

Dedicated channels may be acquired using the novel TBSD-RRM protocol to be introduced in chapter 4. Using actively managed radio resources instead of a single shared random access channel is motivated by the idea of utilizing spectrum as efficiently as possible. Common MAC layer implementations, such as the IEEE 802.11 DCF, hardly operate the wireless channel at its peak nominal capacity [79]. This is particularly true in multihop mesh networks.

A convergence protocol is not required for the dedicated channels. Here, each channel employs an instance of a physical layer, which allows tuning the SDR to a specific frequency, select a specific scrambling code, request a certain radiation pattern (SDMA) etc. Attention must be payed on the physical layer design, so as to minimize the delay from wireless medium to the dedicated channel's MAC (DMAC) entity. Since the DMAC manages a dedicated channel, there is no need for carrier sense, contention, random back-offs, etc. Instead, a switching matrix must be implemented in L2, which allows forwarding of incoming packets that are destined for other stations. Of course, this must be done based on routing information collected at the network layer. Applications with QoS constraints shall indicate their requirements to the transport layer. The network layer (for instance IPv6) is responsible for selecting a network interface to a dedicated channel, establishing the necessary connections, etc.

2.2.6 Switching

With the availability of dedicated links between stations, new potentials with respect to the switching strategy are opened. Assume that the route has been determined by AODV, or any other suitable routing protocol [61, 80–82]. As a consequence, applications with stringent QoS-requirements would request a dedicated link to the next hop en route to their desired target stations. Intermediate stations would also acquire dedicated links to their respective next hops. Then a variety of switching techniques may be thought of.

As a first example, consider plain circuit-switching of the received symbols using a switching fabric, as depicted in figure 2.4. Using dedicated links, a station receives a symbol on one of its reception channels, while it already knows to which flow the particular symbol belongs. Hence, the symbol can be switched to the corresponding output channel for transmission. Depending on the frame structure of physical layer bursts, this would require only a minimal delay in the order of a few time slots (in a TDMA system). While this technique has been patented, there exists no proof of concept, i.e. there is no demonstrator testbed in operation.

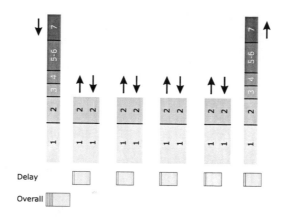

Figure 2.5: L2 Cut-through Switching in Improved MANET

Therefore, a very promising and new hybrid approach is postulated, which combines cut-through and cell switching. Previously, cut-through switching has only been employed in fixed routers with multiple **wired** network links. This is due to the fact that multiple distinct physical channels are required for transmission and reception which required costly radios with lots of analog circuitry. With the advent of affordable software radio technology, cut-through switching in wireless networks becomes a viable option. When we assume a cell structure comparable to that of ATM, only a few header bytes need to be examined in order to retrieve the virtual channel and virtual path identifiers of the cell. After successful decoding of the (short) cell header, forwarding of the payload could start immediately – resulting in a dramatic reduction in end-to-end delay, as illustrated in figure 2.5. In addition, different data flows in a single station may be separated by orthogonal variable spreading factor (OVSF) channelization codes (e.g. WALSH codes), as it is done in UMTS.

2.2.7 Virtual Circuit Establishment

Virtual channels may be obtained by forwarding resource reservation packets on a previously determined route from source to destination. These packets can be sent on the same common control channel also used to deliver RRM management frames, while the routes may be determined by a suitable routing protocol, such as AODV. An example of two overlapping routes is illustrated in figure 2.6. These routes have been determined using the new AODV implementation integrated into the simulation environment, which is going to be presented in part II. But other techniques integrating routing and virtual circuit establishment [83] may be thought of as well.

2.3   Conclusion

Starting with an in-depth analysis of the actions and associated delays involved in the data forwarding process in a mobile ad hoc network, a new architecture has been presented. New potentials in the field of high-performance switching in wireless networks have been pointed out. The quite expensive (in terms of delay) operation of route determination is performed during connection establishment and routing information gathered in L3 is used in L2, in order to reduce latency as far as possible. The promising idea of a hybrid cut-through & cell-switching paradigm in MANETs has been outlined, while further research is required to evaluate the strengths and

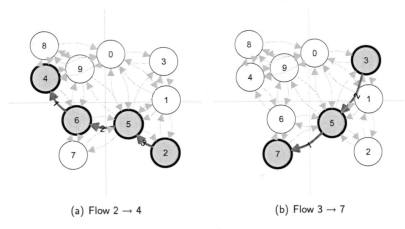

(a) Flow 2 → 4   (b) Flow 3 → 7

Figure 2.6: Two Sample Flows (Routes Determined by AODV)

weaknesses of the proposed architecture and its different flavors. In future work, thorough simulations shall provide insight to throughput/delay characteristics, allowing a comparison against state-of-the-art technology. The presented architecture shall serve as a framework for the various building blocks that are going to be detailed in the following chapters.

This page intentionally left blank.

Part I

New Algorithms and Protocols

This page intentionally left blank.

## CHAPTER 3

Extensions to the IEEE 802.11 MAC

In this short chapter, three enhancements of the standard IEEE 802.11 MAC suitable for a broad range of applications are outlined. The first one is a simple, but efficient, method for improving reliability of multicast frames. It offers instant performance gains for protocols that make use of the WM's inherent broadcast nature to disseminate data efficiently. Originally, its design was motivated during development of the novel protocol suite detailed in chapter 4. But it turned out to be of general interest to a large class of protocols for wireless networks.

As a second extension, two special queues are presented. The first one, a rate limiting queue, provides a means to reduce network congestion using open-loop control. The other one, a priority queue, is able to improve protocol performance in presence of congestion. Messages are reordered, such that more urgent messages are delivered sooner than less urgent ones. For instance, responding to a recently received request may be more urgent than issuing a new request. However, this strongly depends on the protocol considered.

The third extension provides accurate round-trip-delay measurements over a random access channel, which is governed by IEEE 802.11 DCF. It circumvents the problem of unpredictable, time-varying delays in transmissions, which would otherwise render delay measurements useless. Such delay measurements may be used for GPS-free localization in multihop ad hoc networks, for example.

3.1 More Reliable Multicasts

Many protocols for wireless networks make use of the medium's inherent broadcast feature to deliver management messages intended for a number of immediate neighbors in an efficient way. In particular, the suite of protocols presented in chapter 4 also belongs to this class. However, there is one problem associated with multicast frames transmitted over IEEE 802.11 MAC, namely that of dramatically reduced reliability compared to unicast frames. This reduction is caused by several issues, which shall be studied in the following.

3.1.1 Existing Mechanisms for Reliable Unicasts

The standard DCF offers three orthogonal mechanisms for reliable transmission of unicast frames: immediate positive ACKs, fragmentation, and a priori medium reservation under RTS/CTS

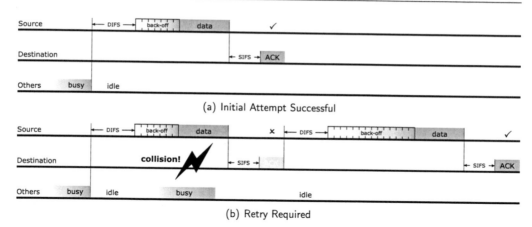

(a) Initial Attempt Successful

(b) Retry Required

Figure 3.1: Acknowledgment in IEEE 802.11 DCF

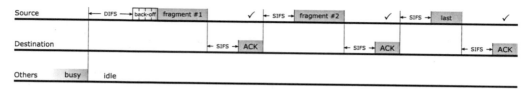

Figure 3.2: Fragment Burst in IEEE 802.11 DCF

regime. The receiver will acknowledge successful reception of any MPDU where it appears as addressee immediately after reception. It matches its own burnt-in hardware address against the address field in the MAC frame header to decide whether it is the addressed station or not. If so, it responds immediately with an ACK frame after a SIFS period – without prior carrier sense, and without respecting the back-off counter value[1].

The source receives the ACK frame and recognizes that transmission must have been successful. This case is illustrated in figure 3.1(a). However, when the source does not receive an ACK frame in time, it assumes that transmission must have failed, performs the random back-off procedure and retransmits the lost packet, as depicted in figure 3.1(b). By the way, it is also possible for the source not to receive an ACK frame in spite of the destination having actually acknowledged successful reception. In this case the destination might receive the same frame twice. Such duplicate frames are easily rejected using a source sequence counter.

Fragmentation, which is the second mechanism, is used to improve probability of successful transmission for long frames. Lengthy data frames are more likely to collide than shorter ones due to the longer time the medium is required to be interference-free. Even worse, the cost of loosing a lengthy fragment is also higher compared to a short fragment, because the medium will be occupied for a longer period of time, when the whole frame is retransmitted. This is where fragmentation comes into operation: the originally long frame is split into several shorter frames, so-called fragments, which are transmitted in a fragment burst. Each short fragment is individually protected by an ACK frame, as shown in figure 3.2. This way, when a fragment happens to be lost, only the short fragment needs retransmission.

Finally, **a priori** medium reservation under RTS/CTS regime is the third mechanism used to protect directed frames. The most important feature of this technique, which has also been

---

[1]Refer to section 1.6.3 for details on basic channel access under DCF regime

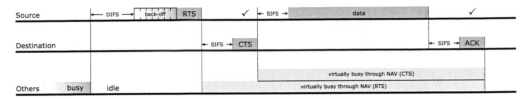

Figure 3.3: Virtual Carrier Sense in IEEE 802.11 DCF

mentioned in the introductory chapter, is its ability to mitigate the hidden terminal problem via virtual carrier sense. This becomes obvious, when we refer to figure 3.3. Here, an RTS frame is transmitted in advance of the actual data frame. The intended destination immediately responds with a CTS frame after a SIFS period. On one hand, this serves as an acknowledgment to the source, indicating successful reception of the short RTS. On the other hand, terminals that were unable to receive the RTS frame might have been able to receive the CTS frame. Both frames, RTS and CTS, indicate the duration of the actual data transfer, which is about to follow. Hence, stations that receive either the RTS frame (i.e. those in vicinity of the source) or the CTS frame (i.e. those near to the destination), will adjust their network allocation vector (NAV) accordingly. The NAV is chosen such that the channel is virtually busy for the duration of the data frame and its corresponding ACK. The RTS/CTS method may also be combined with fragmentation. In this case, the CTS plays the role of the first ACK, seizing the medium for the duration of the first fragment plus ACK. Subsequent ACKs reserve the medium for the duration of a consecutive fragment plus ACK.

3.1.2 Simple, Yet Efficient, Method for More Reliable Multicasts

It is worth mentioning that positive ACKs are used for unicast transmissions only. Broadcast and, more generally, multicast frames are not protected by ACKs, cannot be fragmented into smaller portions and are not supported by the RTS/CTS mechanism. This is because of the well-known ACK explosion problem. Hence, they are exposed to increased loss probability compared to directed frames. For many applications it would be desirable to have a way of distributing information as efficiently as a broadcast does with the reliability that a unicast provides.

Therefore, a very simple – yet crucial – extension to the DCF is proposed to support such applications. The enhanced DCF shall handle an additional class of MAC layer management protocol data units (MMPDUs). Here, third stations are allowed to listen to such MMPDUs albeit not being the addressed receiver. That is, these intentionally broadcast MMPDUs are being sent as directed unicast transmissions. The only difference is that receivers deliver the frame even if the receiver address specified in the frame header does not match the local hardware address. It is up to client protocols to examine the receiver hardware address, if required. This way, fragmentation, RTS/CTS and positive ACK techniques are immediately made available to such frames at a very little cost (in fact the same as for any other unicast frame). While many stations listen to this kind of transmission, only a single station is immediately addressed and thus only this very station will respond with CTS and ACK frames. Therefore, reliability is greatly improved, while efficiency does not suffer noticeably.

This new class of more reliable management broadcasts has successfully been used in the development of the NDP and TBSD-RRM protocols to be discussed later on. For example, the TBSD-RRM protocol uses such broadcasts with enhanced reliability to spread information effectively and reliably. For each broadcast, it randomly picks one of its immediate neighbors as the addressee. This ensures that at least one neighbor will receive the message with utmost probability. And if it does not, TBSD-RRM will know that delivery has failed thanks to the missing ACK response. It can then decide to pick another neighbor or take completely different action.

As a result, this modified version of 802.11 DCF is perfectly suited as a means for asynchronous exchange of signalling PDUs among stations of next-generation wireless ad hoc networks. While it does not guarantee that all neighbors receive the broadcast transmission, at least one neighbor will receive the transmission with high probability, and at low cost (in terms of overhead). Randomly picking this neighbor from the set of immediate neighbors at each round, further contributes to the "survival" of important information, for example a global state. Hence, this slight but effective enhancement has been presented as a self-contained topic, so as to promote its application to a variety of existing and new protocols in the sector. For example, this technique can be effectively combined with a novel strategy for efficient, yet reliable, transmission of sensor values over wireless networks, which is going to be detailed in chapter 5. Furthermore, the ARP protocol can instantly benefit from this novel strategy, as it is going to be argued in section 6.8.2.

3.2 Priority and Rate Limiting Queues

Under high network load, the DCF's TX queue might grow, because the channel does not become available to a particular station for some amount of time. In this case, it is not reasonable to push more and more packets onto the queue in an uncontrolled fashion, since this would cause severe congestion in the network. In other words, under heavy network load, the TX FIFO queue would grow and grow, without the ability to drop less important packets, influence the actual transmission order, or take similar action. To improve the situation, the DCF has been enhanced by two concepts: a rate limiting queue, which limits the number of frames forwarded to the actual TX queue; and a priority queue, which favors certain types of frames over others. In general, both concepts could be combined.

3.2.1 Rate Limiting Queue

The rate limiting queue limits the number of MSDUs per time unit it forwards to the actual TX queue. For example, the limit may be set, such that no more than m MSDUs per second are allowed to be passed to the TX queue. When the first MSDU is pushed onto the rate limiting queue, it is immediately forwarded to the TX queue, and the current time is memorized. Then, the next MSDU that is eventually pushed on the queue is not dequeued before $1/m$ seconds have elapsed. Compared to priority queues, the order of MSDUs is not modified, and the rate limiting queue does neither need nor utilize feedback from the DCF. One could say that the rate limiting queue implements open-loop control.

3.2.2 Priority Queue

The priority queue reorders pending transmissions, such that those with higher priority are dequeued first. In addition, it uses feedback provided by MA-UNITDATA-STATUS.indications to determine when to dequeue the next pending MSDU. Compared to the rate limiting queue, the control loop is closed.

Initially, when the queue is empty, the first MSDU pushed onto the queue is immediately forwarded to the MAC's TX queue, where it is stored until the channel becomes available. However, it also remains on top of the priority queue, until the MAC service has reported either

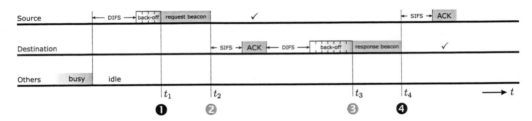

Figure 3.4: Proposed Operation of DMF

successful transmission or a failure condition (retry limit exceeded, lifetime expired, etc.) through its MA-UNITDATA-STATUS.indication. This is the time, when the packet is taken out of the queue and the next pending packet is forwarded to the MAC service for delivery. This is necessary, because a packet which has once reached the MAC service's TX FIFO queue is out of hand.

3.3 Accurate Round-Trip-Delay Measurement Function

Sometimes it is useful to estimate the distance between two distinct stations. For example, using a set of distance measurements it is possible to infer the underlying geometry, that is, the network topology. This problem is well-known as the molecular conformation problem [84]. One way to determine the distance is to estimate the time it takes an electromagnetic or ultrasonic wavefront to travel from source to destination. Let us assume that the medium is the air. Thus, velocity of an electromagnetic wave is approximately equal to speed-of-light, c. The EUCLIDean distance d may then be determined by measuring air propagation time τ, according to $d = \tau c$. A powerful means to come up with accurate delay estimations is pilot-signal assisted channel-estimation, for example.

3.3.1 A Novel Delay Measurement Function

Some systems, however, may not employ channel estimation, or are not designed to indicate current estimations to interested client protocols. In these cases, round-trip-delay measurement may be a viable option. In addition, round-trip-delay measurement may be used as a complement, in systems where channel estimation information is in fact available. Here, the delay is measured by observing the time a data packet needs to travel from source to destination. This time is obviously not equal to air propagation time due to the associated processing overhead, but it is possible to estimate propagation time, when these overheads are determinable.

A problem arises with packets transmitted under control of the DCF: carrier sense and random back-off times are unpredictable. Therefore, a novel Delay Measurement Function (DMF) is proposed as an extension to the standard DCF. Similar to the IEEE 802.11 Timer Synchronization Function (TSF), time stamps are used to compensate the effects of stochastic delays. The DMF's proposed modus of operation is outlined in figure 3.4.

Suppose that a client protocol is interested in a delay measurement to one of its immediate neighbors. The DMF would put a request beacon MPDU into the MAC's TX queue. An MPDU, once put onto the queue, is out of hand. To be more specific, neither can it be easily removed from the queue, nor does the source know, at what instant in time the MPDU is actually put "on-the-air". This issue is resolved by the proposed extensions to the DCF. The DMF request beacon contains a time stamp which identifies the start of actual transmission in terms of source station timer ticks. In the following paragraphs, $c_s(t)$ shall denote the source station clock count at time t, whereas $c_d(t)$ shall denote destination station timer ticks. Naturally, clock counts are integer numbers, i.e. $c_{d/r} : \mathbb{R} \mapsto \mathbb{N}$.

The extended DCF shall update this time stamp immediately before the first bit is passed to the PHY for transmission. This point in time, t_1, is marked as (1) in figure 3.4. This is when the DCF will know that transmission is about to start, thanks to the `PHY-TXSTART.confirm` it receives from the IEEE 802.11 PHY. The destination recognizes the incoming data frame, as soon as it is alerted by a `PHY-RXSTART.indication`. However, it does not know what kind of frame it has started to receive, until the last bit has been transferred from PHY to MAC. At this point (2) the destination saves its local timer count $c_d(t_2)$ for later use. As for any directed frame, proper reception is acknowledged with the help of an immediate ACK.

Now, the destination prepares a response frame, which it puts onto its MAC TX queue. Again, once in the TX queue, the response beacon MPDU is out of control. The medium may be seized by other stations, the queue may be congested by several other MPDUs pending for transmission, etc. Eventually, the medium will have been idle for more than a DIFS plus the destination's current back-off duration. As soon as transmission is about to start (3), this event is again reported to the MAC by a `PHY-TXSTART.confirm` from the PHY. This is also the time, when the destination calculates the amount of time elapsed since it received the initial request beacon. It does so with the help of the previously saved value $c_d(t_2)$

$$\varepsilon = c_d(t_3) - c_d(t_2).\tag{3.1}$$

It updates the response beacon, storing the calculated amount ε as a "remote processing delay", which can be further evaluated by the source.

When the source has finished reception of the response beacon (4), it calculates air propagation time as half of: its current timer value minus the timer value corresponding to the time when it sent the request (1), minus the remote processing delay (as calculated by the destination), minus the time it required the PHY to transmit request and response beacons, τ_s and τ_d, respectively

$$\hat{\tau} = \frac{1}{2}\left(c_s(t_4) - c_s(t_1) - \varepsilon\right) - \tau_s - \tau_d.\tag{3.2}$$

The latter obviously depends on the data rate, depth of convolutional codes, length of synchronization preambles, etc. However, for any given PHY setup, these times are deterministic and can be obtained through a `PLME-TXTIME.request/indication`.

As a matter of fact, accuracy of the delay measurement mainly depends on timer resolution, which, in turn, depends on the oscillator frequency used to clock the counters. When we assume a 10 MHz timer to measure the propagation delay, a resolution of roughly 15 m can be achieved, while an 80 MHz timer would provide approximately 2 m accuracy. It is worth noting that source and destination timers need not be synchronized, because the DMF works with time differences, and hence, absolute offsets are not critical. It should also be noted that these accuracy considerations are meant as a coarse indicator only, as they do not take performance degradation due to delay spread into account, for example. The achieved accuracy seems to suffice for many applications, like localization/positioning. An overview of suitable positioning techniques is provided in [85].

## CHAPTER 4

An Innovative Approach to
Distributed Dynamic Channel Assignment

This chapter is concerned with a novel fully distributed radio resource management (RRM) protocol, which offers collision-free physical channel assignments for virtually any kind of wireless network. In particular, it may be used to manage dedicated peer-to-peer links in ad hoc networks. In this sense, the proposed protocol serves as an enabling technology for novel switching concepts in MANETS, for instance L2 cut-through switching[1]. Thanks to the flexible and generic approach inherent to the presented solution, the new protocol is also useful for load-adaptive resource management in 2G/3G networks, for instance. It also simplifies a graceful transition to coverage extension around hotspots via multihop technology for the same reason.

In order to tackle the channel assignment task in a distributed fashion, a modular suite of appropriate protocols has been designed. The modular structure facilitates reuse of certain portions with general applicability. Moreover, it helps to improve certain isolated parts in the future, while others remain unchanged. Hence, modularity facilitates maintenance of the novel protocol suite and simplifies integration of future enhancements.

On the top level, two protocols have been designed, extending the basic functionality of the classic IEEE 802.11 MLME. The first one is responsible for gathering neighborhood topology information, and is consequently called Neighborhood Discovery Protocol, or NDP in short. The other one exploits information provided by NDP and performs the actual resource management task. It is called transaction-based soft-decision radio resource management protocol, and is abbreviated TBSD-RRM.

In the chapter's introductory part, the RRM protocol's theoretical foundation – UxDMA, one of the best-performing **centralized** dynamic channel assignment (DCA) algorithms known to date – is briefly reviewed. Afterwards, the context in which the new protocol suite shall operate is going to be highlighted. Subsequently, the individual modules that make up the entire protocol suite are introduced and discussed. Finally, the assignments achieved by the innovative, fully distributed regime are compared against the approved, centralized algorithm for a number of representative network topologies. Preliminary results have been published in [86, 87].

---

[1]Refer to chapter 2 for details of this concept

4.1   Introduction

Most wireless ad hoc networks employ some flavor of carrier-sense multiple access, for instance CSMA/CA, in order to establish a contention-based channel access scheme. Its appeal lies in the simplicity and relative ease with which a fully distributed regime can be implemented. Unfortunately, the price to pay is degraded network performance in terms of throughput/delay characteristics, due to near-far phenomena (i.e. the well-known hidden and exposed terminal problems), fading and capture effects on the channel. In addition, QoS and fairness issues are generally difficult to address with such kind of channel access schemes, pointing the way to planned/scheduled MAC schemes.

Planned or scheduled MAC protocols prearrange or negotiate a set of time slots, frequency bands or scrambling codes for individual stations or links, such that transmissions from these nodes or on these links do not collide with other scheduled transmissions within interference range. TDMA, FDMA, CDMA, SDMA, and combinations thereof are widely deployed and approved in 2G/3G cellular systems. Here, a central coordinator (base station) performs intra-cell resource management, while inter-cell management is often done by the network operator using static strategies, e.g. fixed channel assignment (FCA) [88], in order to facilitate spatial reuse. Obviously, this approach is difficult to apply to self-organizing structures due to the absence of base stations and a network operator.

4.2   Previous Work and Theoretical Foundation

In previous work, a framework together with a centralized algorithm for dynamic channel assignment in wireless networks has been presented by RAMANATHAN [89]. It provides solutions to 144 contemporary and potential channel assignment problems based on a unified approach. Ability to manage such a broad range of problems is founded on the framework in whose context the algorithm operates. The algorithm presents a major contribution, compared to algorithms proposed earlier.

4.2.1   Framework

The physical network topology is transformed into a graph $G = (V, E)$, where V is a set of vertices corresponding to the stations comprising the network and E is a set of **directed** edges between vertices denoting inter-station wireless links. Considering two vertices $u, v \in V$, edge $(u \rightarrow v) \in E$ holds true if v can receive u's transmission[2]. Consequently, the task of assigning channels to stations turns into different flavors of the well-known graph coloring problem, which often turn out to be NP-complete.

Besides the graphical representation modelling the network topology, the specific assignment problem is covered in terms of constraints. A constraint is a symmetric relation between two vertices or two edges. In a legal assignment, two vertices/edges that are mutually constraint must receive different colors. Thus, a constraint imposes a restriction on coloring. Eleven atomic constraints – four in terms of vertices, and seven in terms of edges – underlying most assignment problems offer a means to express various problems as a combination thereof. Figure 4.1 illustrates these constraints, where the black vertices/edges are mutually constrained, while the gray ones cause the constraint.

---

[2]Notice that $(u \rightarrow v) \in E$ does not imply $(v \rightarrow u) \in E$

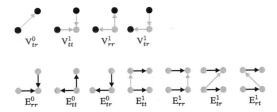

Figure 4.1: Eleven Atomic Constraints

Problem	Constraint Set
Cellular frequency assignment	$C = \left\{ V_{tr}^0 \right\}$
TOCA/ROCA CDMA code assignment	$C = \left\{ V_{tt}^1 \right\}$
TDMA/FDMA broadcast schedule/assignment	$C = \left\{ V_{tr}^0, V_{tt}^1 \right\}$
POCA CDMA code assignment	$C = \left\{ E_{rr}^0, E_{tr}^0, E_{tt}^0 \right\}$
TDMA/FDMA link schedule/assignment	$C = \left\{ E_{rr}^0, E_{tt}^0, E_{tr}^0, E_{tr}^1 \right\}$
Full duplex TDMA/FDMA link schedule/assignment	$C = \left\{ E_{rr}^0, E_{tt}^0, E_{tr}^1 \right\}$
TDMA/FDMA/SDMA schedule/assignment	$C = \left\{ E_{rr}^0, E_{tr}^0 \right\}$

Table 4.1: Constraint Sets for Popular Assignment Problems

Table 4.1 maps some popular assignment problems to their according constraint set representation. Let us consider TDMA broadcast schedule as an example. This kind of problem arises in wireless multihop networks (packet broadcast), when a station's transmission is intended for all neighboring stations. Consequently, if a station is assigned a timeslot and/or frequency band then none of its out-neighbors should be assigned the same resource since they will have to tune their receivers on the specific timeslot/frequency ($\to V_{tr}^0$ constraint). To avoid collisions at the out-neighbors, each of their in-neighbors must also not be assigned the same physical resource in turn ($\to V_{tt}^1$ constraint). Hence, this problem is characterized by $C = \left\{ V_{tr}^0, V_{tt}^1 \right\}$. The outcome of a valid coloring is visualized for a certain node distribution in figure 4.5 on page 47.

4.2.2 Centralized Assignment Algorithm: UxDMA

Running the algorithm UxDMA (listing 4.1) presented by RAMANATHAN on a network graph yields the desired colors or, practically speaking, channel assignments. UxDMA will assign colors to the different network nodes or links between nodes subject to the specified constraint set C and ordering heuristic. The algorithm consists of two phases:

1. **Labelling Phase.** During labelling, each vertex is assigned a unique label $l = 1 \dots |V|$, which will determine the processing order later applied in the coloring phase. Labels are comparable to priorities, i.e. a vertex with a greater label is considered more important than one with a smaller label. In other words, vertices with higher valued labels are conceded higher priority[3] in the scope of the following phase.

2. **Coloring Phase.** Here, vertices are considered in the order determined during labelling. In case of edge coloring, the edges incident on the considered vertex v are colored, in case of vertex coloring, v itself is colored. The color is chosen in a greedy fashion, i.e. the first available color is assigned, such that none of the constraints are violated.

---

[3]One might also say importance, relevance or weight

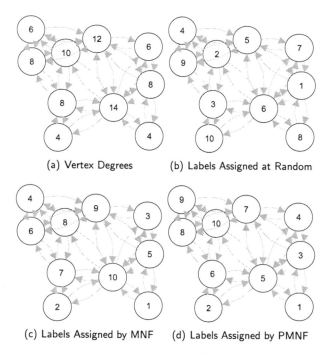

(a) Vertex Degrees   (b) Labels Assigned at Random

(c) Labels Assigned by MNF   (d) Labels Assigned by PMNF

Figure 4.2: Label Assignments Obtained by Different Heuristics

The ordering heuristic mentioned above affects the way the labelling is done, i.e. it determines the priority of each vertex during the coloring phase. In figure 4.2, the results of the labelling process are depicted for a simple network consisting of ten stations. Three ordering heuristics have been presented by RAMANATHAN, while generally others might be thought of as well:

- **Random.** The simplest ordering, vertices are colored at random.

- **Minimum Neighbors First (MNF).** Here, vertices that have fewer neighbors are assigned smaller numbers. As the coloring is done in decreasing order of labels, high-density vertices are colored first.

- **Progressive Minimum Neighbors First (PMNF).** This heuristic is similar to the previous one, with a subtle (yet crucial) difference: once a vertex v has been labeled, v and all edges incident on v are ignored for the remainder of the labelling phase. Therefore, the neighborhood of a vertex keeps on changing, while other vertices are labeled.

Notice that the random assignment shown in 4.2(b) is just a potential realization, i.e. consecutive assignment runs will produce significantly differing labellings with high probability. Similarly, MNF and PMNF might produce slightly differing assignments in repetitive runs, when there are a number of vertices with equal degrees to choose from during the labelling phase.

Covering the algorithm in detail is far beyond the scope of this work. However, it should be noted that algorithm UxDMA is remarkable for several reasons. The algorithm operates on arbitrary graphs, i.e. neither does it assume bidirectional links, nor does it require the network graph to be restricted in any form, e.g. to be a tree, planar graph, disc graph, etc.

In particular, for most problem classes, the PMNF ordering results in colorings with tight worst-case (theory) and in-practice (simulation) performance bounds with respect to the optimum

algorithm UxDMA

input: (1) directed graph $G = (V, E)$
       (2) element type $\Psi \in \{\text{vertex}, \text{edge}\}$ to be colored (assigned)
       (3) constraint set C which is either
           $C \subseteq \{V_{tr}^0, V_{tt}^1, V_{tr}^1, V_{rr}^1\}$ (when $\Psi = \text{vertex}$), or
           $C \subseteq \{E_{tt}^0, E_{tr}^0, E_{rr}^0, E_{tt}^1, E_{tr}^1, E_{rt}^1, E_{rr}^1\}$ (when $\Psi = \text{edge}$)
       (4) ordering $\omega \in \{\text{random}, \text{mnf}, \text{pmnf}\}$

output: A $C-$constrained assignment of q colors 0 through $q-1$
        to vertices (when $\Psi = \text{vertex}$) or edges (when $\Psi = \text{edge}$) of G

begin
    for all elements e of type Ψ **do**
        $\text{color}(e) \leftarrow 0$
    Assign-Label (G, ω)
    for j from largest down to smallest label **do**
        let $u \leftarrow$ vertex with label j
        for each element e in Surround (G, Ψ, u) **do**
           $\text{color}(e) \leftarrow$ First-Available-Color (G, e, C)
end

procedure Assign-Label (G, ω)
    for labels l from 1 through $|V|$ **do**
        if $\omega = \text{random}$ **then**
           pick unlabelled vertex $u \in V$ at random
           $\text{label}(u) \leftarrow l$
        else if $\omega = \text{mnf}$ **then**
           pick unlabelled vertex $u \in V$ of minimum neighbors in G
           $\text{label}(u) \leftarrow l$
        else if $\omega = \text{pmnf}$ **then**
           pick unlabelled vertex $u \in V$ of minimum neighbors in G
           $\text{label}(u) \leftarrow l$
           delete all edges incident on u from G
return

function Surround (G, Ψ, u)
    if $\Psi = \text{vertex}$ **then**
        $S \leftarrow u$
    else
        $S \leftarrow$ set of incoming and outgoing edges incident on u in G
return uncolored subset of S

function First-Available-Color (G, e, C)
    $taken \leftarrow \emptyset$
    for each constraint $c \in C$ **do**
        for each colored element f of type same as e such that e, f are related by constraint c **do**
           $taken \leftarrow taken \cup \text{color}(f)$
return smallest color $\notin taken$

Listing 4.1: Formal Specification of UxDMA

colorings. Specifically, this heuristic achieves colorings that are guaranteed to be within $O\left(\theta\right)$ of the optimum colorings, where θ is the topology graph's thickness[4]. Algorithms proposed earlier, were only able to guarantee performance bounds of $O\left(\rho\right)$, where ρ denotes the maximum graph degree[5]. Comparing both algorithm classes, solutions with guarantees proportional to θ are clearly more scalable, because θ has been shown to be several orders of magnitude less than ρ for typical wireless multihop networks. In fact, PMNF has been observed to provide an in-practice guarantee of $1.1\times$ optimum.

In figure 4.3 a network consisting of 100 vertices and 484 edges is illustrated. Here, $C = \left\{\mathrm{E}_{rr}^{0}, \mathrm{E}_{tr}^{0}, \mathrm{E}_{tt}^{0}, \mathrm{E}_{tt}^{1}, \mathrm{E}_{rr}^{1}, \mathrm{E}_{tr}^{1}, \mathrm{E}_{rt}^{1}\right\}$ has been specified and the PMNF heuristic has been used to produce the colorings which are printed on the arrows representing the edges. In total, 80 colors have been assigned to the 484 edges. In figure 4.4, the spatial reuse pattern for the same colorings is visualized. This is done by printing those edges in bold that have been assigned the first (and most often used) color $c_0 = 0$, which in this case has been assigned to 20 distinct edges. Both figures give an impression of the constraint set's impact on the colorings.

Consider figure 4.5 as a further assignment example. Contrary to the previous one, vertices (not edges) are the elements to be colored, subject to $C = \left\{\mathrm{V}_{tr}^{0}, \mathrm{V}_{tt}^{1}\right\}$. Again, vertices that have received the assignment $c_0 = 0$ are highlighted, so as to visualize the reuse pattern. In a TDMA or FDMA system, highlighted stations would be able to transmit concurrently on the same physical channel, i.e. using the same carrier frequency and/or time slot – effectively avoiding mutual interference.

4.3 Requirements for a Real-World Solution

4.3.1 Limited View of Network Topology

The algorithm presented in the previous section is not immediately applicable to wireless multihop networks, because it operates in a centralized fashion, i.e. it requires complete knowledge of the network topology. In a pure wireless network with thousands of stations, gathering and maintaining such kind of information at a central coordinator is by no means practicable: It would take tremendous time to build the network graph, let the central coordinator assign the channels and then distribute the assignments back throughout the entire network. Such a procedure might only be acceptable when dealing with static assignment problems, where resources are assigned once during initial network startup with very seldom changes in network topology.

However, in a flat wireless multihop network topology changes quite frequently due to mobility, fading effects, power-source failures, stations joining the network after initial startup etc. In such a scenario, it is simply not feasible to perform assignments in a centralized way. Instead, a fully distributed regime is required, meaning that the assignment task is tackled jointly by all stations that belong to the network (or a suitable subset thereof). To achieve the common goal of channel assignment, several instances of a proper algorithm are run on participating stations, using message exchanges to coordinate their efforts. These individual instances must be able to get by with a limited view of the network topology, where the limitation may be quantified in terms of hops. An algorithm that requires knowledge of its immediate neighbors only clearly scales better than a competing algorithm in need of two-hop topology information.

---

[4]A graph's thickness θ equals the minimum number of planar graphs into which it can be partitioned. In other words, thickness yields a measure of "nearness-to-planarity".

[5]The maximum degree of a graph is the maximum of the total degrees taken over all vertices of the graph, where the total degree $d_G\left(v\right)$ of a vertex v equals the number of edges (incoming and outgoing) incident on v, i.e. $d_G\left(v\right) = \left|\left\{w\mid \left(v \rightarrow w\right) \in E \vee \left(w \rightarrow v\right) \in E\right\}\right|$.

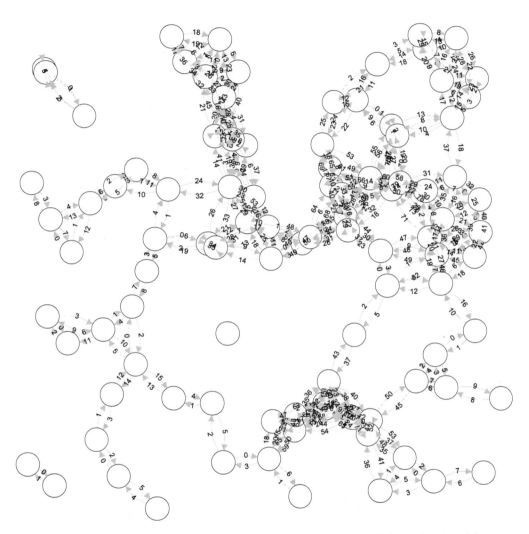

Figure 4.3: Colorings Produced by PMNF for $C = \left\{ E_{rr}^0, E_{tr}^0, E_{tt}^0, E_{tt}^1, E_{rr}^1, E_{tr}^1, E_{rt}^1 \right\}$

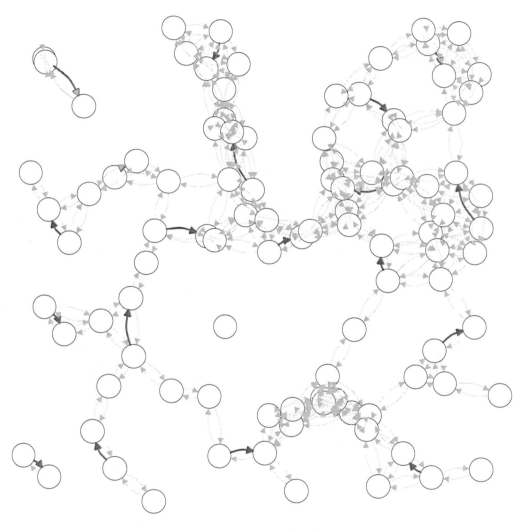

Figure 4.4: Reuse Pattern for $C = \left\{ \mathrm{E}^0_{rr}, \mathrm{E}^0_{tr}, \mathrm{E}^0_{tt}, \mathrm{E}^1_{tt}, \mathrm{E}^1_{rr}, \mathrm{E}^1_{tr}, \mathrm{E}^1_{rt} \right\}$

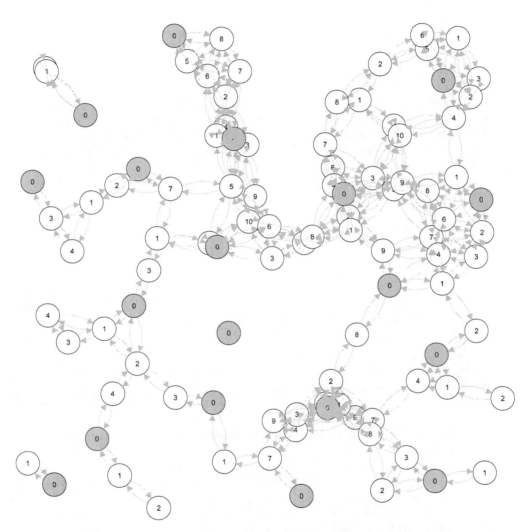

Figure 4.5: Colorings and Reuse Pattern for $C = \left\{ V_{tr}^0, V_{tt}^1 \right\}$

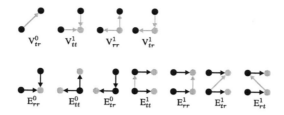

Figure 4.6: Managing Vertices in a Distributed Version

Requirements Analysis

Let us consider the constraints mentioned in the previous section in order to determine the neighborhood information depth required to decide whether a certain constraint is met or not. In case of the vertex constraints, one-hop topology information is sufficient to evaluate V_{tr}^0, i.e. each station must know its immediate neighbors. The remaining three vertex constraints, V_{tr}^1, V_{tt}^1, and V_{rr}^1 need two-hop topology information.

The situation is a bit different with edge constraints, though. Despite edges being the conceptual entities in need of channel assignment, only stations might be the acting elements in any distributed system. In other words, stations (vertices) have to initiate and manage channel assignment in a distributed system. In the scope of this work, the following convention shall apply:

A conceptual edge $e = (u \rightarrow v)$, i.e. an inter-station link between stations u and v, is **managed** by u, that is, the associated **transmitting vertex**. In other words, the so-called **managing vertex** u carries out all actions involved in the channel assignment process **on behalf of** e.

Managing vertices for said constraints are depicted in figure 4.6. In contrast to the coloring scheme used earlier in figure 4.1, black vertices now mark managing entities. Notice that for vertex constraints, vertices themselves remain the managing entities. Bearing these conventions in mind, let us analyze the requirements with respect to neighborhood information depth. First of all, we can observe that there are generally two managing vertices for each constraint – except for E_{tt}^0, where both edges are managed by the same vertex.

It can be easily seen that E_{tt}^0 demands for one-hop neighborhood information **around the managing vertex**. Similarly, the E_{rr}^0 constraint requires two-hop topology information, regardless of the managing vertex considered. To complete the analysis for the distance-0 edge constraints, E_{tr}^0 demands for one-hop or two-hop neighborhood knowledge, respectively. Hence, two-hop neighborhood knowledge around each managing vertex is in general critical for successful constraint evaluation.

Following the same approach, we can identify the critical depths for the remaining distance-1 edge constraints. While E_{tt}^1 always needs two-hop information and E_{rr}^1 always needs three-hop information, E_{tr}^1 and E_{rt}^1 either need two- or three-hop information, depending on the particular edge taken into account. Again, three-hop deep topology information ensures a legal assignment under all circumstances. When we consider managing vertex separation and required topology information depth as problem complexity measures, six problem classes may be identified, as summarized in table 4.2.

Constraint	V_{tr}^0	V_{tt}^1	V_{tr}^1	V_{rr}^1	E_{tt}^0	E_{tr}^0	E_{rr}^0	E_{tt}^1	E_{tr}^1	E_{rt}^1	E_{rr}^1
Problem Class (Complexity)	II	IV	IV	IV	I	III	IV	III	V	V	VI
Managing Vertex Separation	1	2	2	2	0	1	2	1	2	2	3
Topology Information Depth	1	2	2	2	1	2	2	2	3	3	3

Table 4.2: Managing Vertex Separations and Critical Depths for Various Constraints

4.3.2 Message-Passing Facility

As already mentioned above, individual instances of a distributed algorithm rely on a message-exchange facility to coordinate their joint efforts. We must assume, that control messages are passed to peer entities on a single shared channel with contention-based MAC (like ALOHA, slotted ALOHA, CSMA/CA, etc.). In any case, messages are transmitted at limited data rates, meaning that a transmission cannot occur in "no time".

Furthermore, since our objective is to construct a conflict-free schedule, we must assume that no such schedule exists, i.e. packets may collide on the channel leading to a complete message loss. Even if messages are protected by immediate positive ACKs and CRC/ARQ schemes in L2 (as in IEEE 802.11), retry limits may have been exceeded or lifetime counters may have expired. In addition, the medium may be blocked by on-going transmissions, resulting in unpredictable message delays and growing packet queues. Furthermore, it is generally difficult to protect broadcast and multicast frames due to the well-known ACK explosion problem.

This leads to further problems regarding the message sequence: messages originated at different transmitters at the same time will be processed by their respective intended receivers either simultaneously (when transmissions have sufficient spatial separation), or in arbitrary sequential order, depending on the traffic conditions in the respective areas. These considerations have to be taken into account, when designing a "real-life" protocol – otherwise, dead-locks, loops, etc. become unavoidable.

In some cases, token passing [90] might be an appropriate answer to some of these problems, in the sense that stations have to contend for a token, where only a single station at any instant in time might own the token. Then, the one and only station in possession of the token is allowed to take some particular action. However, token generation, token passing and token loss recovery present problems of their own in the harsh wireless multihop environment.

4.4 Neighborhood Discovery Protocol

Many algorithms related to wireless multihop networks require some sort of topology information, exactly like the TBSD-RRM protocol to be described later on. Therefore, a generic protocol, which integrates into the MLME, has been developed as part of this work, providing such kind of information to its clients.

4.4.1 Local Neighborhood Tables

Individual NDP instances maintain a table of known connections (links, edges) within a limited scope of their neighborhood. A table entry related to the edge $(u \rightarrow v)$ contains u's and v's hardware address, a distance indication $d \in \mathbb{N}$, and a lifetime for the link. Here, d keeps track of the distance to the station, which has reported existence of the particular edge. For instance, if station u itself has detected a link to one of its immediate neighbors v, the associated distance for $(u \rightarrow v)$ is set to $d = 0$.

The lifetime ensures that edges related to stations, which have failed or moved out of range, are removed from the table. If an entry's lifetime has expired, the associated edge is considered deprecated. Lifetime counters are prolonged, whenever a "fresh" indication of link existence is available. Link existence may be reported by peer NDP instances or detected by the station itself using the mechanism described in the following paragraphs. In addition, links to immediate neighbors may be constantly monitored by evaluating the statistics collected in the IEEE 802.11 MAC, anyway. For instance, when a packet could not be delivered even after the maximally allowed number of retries, it can be assumed that the link to this station is lost.

4.4.2   Probe Requests

When a station is powered-on, an SME-POWER.indication notifies NDP of the power-on condition. In response to this indication, NDP schedules an initial "hello!" message, or more formally, an NDP-PROBE.request, which shall be broadcast within the next τ seconds, where

$$\tau = r \cdot \sigma \qquad r, \sigma \in \mathbb{R}. \tag{4.1}$$

In this case, σ specifies the so-called spread, a configurable system parameter, whereas r denotes the realization of a pseudo-random variable R drawn from the standard uniform distribution, i.e. with a probability density function according to

$$f_R(r) = \begin{cases} 1 & \text{for } 0 \le r \le 1 \\ 0 & \text{otherwise} \end{cases}. \tag{4.2}$$

Since the station has originally got no information about its neighborhood at startup, the initial message contains no useful information except the station's own hardware address. The message is sent as a management broadcast frame from within the (thereby extended) MLME. The same request frame is repeatedly transmitted up to $m \in \mathbb{N}$ times in a row, in order to increase probability of successful transmission. This is necessary, because broadcasts are not protected by L2 ACK frames. Moreover, probe requests are transmitted periodically, to cope with node mobility, i.e. a sequence of m probe requests is broadcast every $T \in \mathbb{R}$ seconds. Stations that receive a probe request – may it be an initial "hello!" message or a probe request containing useful information (such as described later on) – shall act in the following way:

1. Assume that u has received a packet from v, then the local NDP table is grown by a new entry for $(u \rightarrow v)$, where the hardware address of v is taken from the MAC frame header's source address field, and $d = 0$. If an entry already exists for $(u \rightarrow v)$ then distance and lifetime information is updated, instead. In general, table entries are only updated with new information if the candidate's distance is less than, or equal to, the existing entry's distance.

2. If the packet contains useful information in form of a remote NDP table, then this information is merged into the local table according to the rules mentioned above. However, the distance field for each entry of the received table is increased by one, before the update is performed.

3. When new entries have been added to the table, all clients of the NDP protocol receive an NDP-NEWCONNECTIONS.indication, with a list of new connections[6]. Also, the next (periodic) probe request frame is scheduled to take place within

$$\tau = T_2 + r \cdot \sigma, \qquad T_2 \in \mathbb{R}, T_2 \ll T, \tag{4.3}$$

---

[6]Notice that a mating NDP-LOSTCONNECTIONS.indication informs clients of lost connections

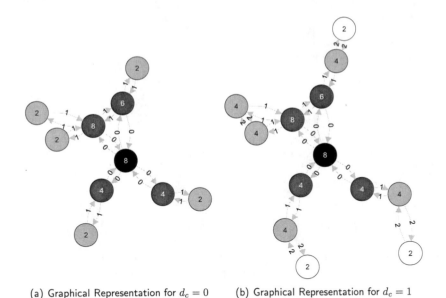

(a) Graphical Representation for $d_c = 0$ (b) Graphical Representation for $d_c = 1$

Figure 4.7: Resulting NDP Graphs (Example)

seconds. In other words, the time span until the next broadcast is shortened.

4. If the destination address specified in the MAC frame header matches the local station's hardware address (or is equal to the broadcast address), an NDP-PROBE.response is issued.

Notice that in addition to the empty probe request frames sent at startup, requests sent while neighborhood discovery is in progress may contain a copy of the requesting station's local table. That is, a station whose local table is not empty, piggybacks this information in the probe request beacon, thereby sharing its local network view with immediate neighbors. The same distance-based table limiting occurs as for response frames (described below).

4.4.3 Probe Responses

A probe response contains a copy of the local station's RRM table, including those elements that satisfy $d \leq d_c, d_c \in \mathbb{N}$. The effect of d_c on the resulting graphs is depicted in figure 4.7. In contrast to probe requests, responses are not sent with the broadcast destination address, but are addressed to the originator of the request. In other words, the concept introduced in section 3.1 is used to distribute the information efficiently and reliably, as it lets probe response frames benefit from L2 ACKs while anyone is allowed to listen to the transmission and extract relevant information. A station receiving a probe response frame acts in the same way, as it does when it receives a probe request beacon. The only difference is that it does not issue a probe response frame.

4.4.4 Parametrization

NDP may be adapted to various network densities, underlying physical channel data rates, etc. using the four different system parameters mentioned earlier:

1. The number of consecutive table broadcasts, m, may be specified.

2. The time diversity within a single round (random inter-broadcast delay) may be influenced by specifying its upper bound σ.

3. The gap between rounds may be specified by T.

4. The cut-off depth d_c for elements included in table broadcasts may be specified.

Numerous simulation runs have proven NDP's ability to provide a local view of the network for many network topologies. Here, $m = 3$, $\sigma = 0.5\,\text{sec}$, $T = 60.0\,\text{sec}$ and $T_2 = 4.0\,\text{sec}$ have turned out as good settings for $d_c = 1$. These settings have been used to obtain the results for all simulations presented in this work.

4.5   Radio Resource Management Protocol

Now that we have local topology information provided by NDP, and UxDMA as a means to produce legal color assignments on these individual **local** network graphs, we lack the glue that fuses the patchwork of colored graphs in such a way that assignments still remain valid in the **global** view. The RRM protocol detailed in this section bridges exactly this gap. It comprises several subsystems (modules), which employ a number of basic concepts frequently used in distributed systems. The outcome is a novel hybrid approach giving an answer to the exceptional challenges imposed by large scale wireless multihop networks[7].

The novelty in the proposed approach lies in the fact that for the first time, a distributed version of the very flexible constraint set based approach is presented, which makes use of a slightly modified version of UxDMA to color local network graphs obtained by NDP. It inherits a number of advantageous properties from its centralized counterpart (mainly flexibility and performance), while it does not require complete topology information, which would render it unusable for highly volatile MANETs. To the author's best knowledge, other proposals published before (e.g. [91–94]) only address specific problems, and do not cope with the potential loss of control messages.

4.5.1   Properties

The proposed RRM protocol offers several advantageous properties, which are going to be discussed in more detail in the course of this chapter:

- It is fully distributed,

- it is modular,

- it requires limited topology information only,

- it uses soft-decision metrics to merge local tables,

- it employs distributed transactions to maintain integrity of global states,

- it uses timeouts and retries to cope with message loss,

---

[7]Distributed DCA is discussed in the context of MANETs here. However, the presented concepts are immediately applicable to many other wireless networks (e.g. existing cellular systems) equally well, because meshed multihop wireless networks constitute the most general and challenging manifestation of the problem

- it can be run incrementally, i.e. for desired parts only,

- it does not require network synchronization for its operation.

In order to distinguish the proposed protocol from other protocols, it will be denoted as transaction-based soft-decision RRM, or **TBSD-RRM** in short.

4.5.2 Overview

The modular concept mentioned before comprises several subsystems, an overview of which is given in figure 4.8. Here, "MMPDU" indicates the transmission and/or reception of MMPDUs using the IEEE 802.11 MAC service. Individual modules are detailed in the remainder of this section. Thus, only a brief description of their basic functionality is given at this point.

A station willing to transmit on a dedicated channel must generally initiate a reservation process beforehand, which, in the end, will result in the so-called "activation" of the particular link. Only activated links are guaranteed not to interfere with others of their kind – provided that network topography is stationary. Therefore, transmissions are only allowed on activated links.

Local Topology (NDP)

Local topology information is proved by an instance of NDP. The TBSD-RRM protocol receives NDP-NEWCONNECTIONS.indications and NDP-LOSTCONNECTIONS.indications, whenever NDP recognizes new links, or detects permanent failure of a previously learned link. This eliminates the need for additional lifetime counters associated with RRM table entries.

RRM Table

Similar to NDP's neighborhood table, the RRM table contains entries that keep track of resources (frequency bands, time slots, scrambling codes, etc.) which have been assigned, together with state and soft-decision information, required by table fusion.

Table Fusion

Table fusion is an algorithm that essentially merges two distinct assignment sets obtained by dynamic channel assignment (e.g. UxDMA). Certain soft-decision metrics are used to determine which assignments should prevail in the presence of conflicts.

Transaction-based Voting Service

In order to activate a link, neighbors are asked to "vote" on a particular assignment proposal. Initially, a vote query is sent to all immediate neighbors, which will then cast positive ("affirmed.") or negative ("veto!") votes. The link is activated, as soon as all due votes have been collected and no veto is among these votes. Timeouts and retries are employed to protect transactional queries.

Distribution Service

Once a link has reached the active state, this information is distributed within the local neighborhood, indicating that the particular resource is no longer available. In a similar way, information about deactivated links is also made public by the distribution service.

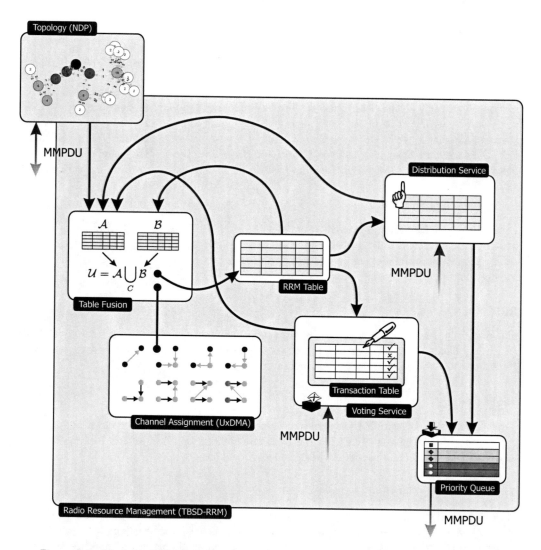

Figure 4.8: Modular RRM Protocol Suite: Subsystems and Their Relations (Overview)

Priority Queue for Management Frames

Under high network load, the MAC layer's transmission queue might grow, because the channel does not become available to a particular station for some amount of time. In this case, it is not reasonable to transmit RRM control frames in the same order, as they where generated. For example, when a number of vote queries are queued for transmission, vote responses should take precedence over pending queries. The priority queue reorders pending transmissions, such that those with higher priority are dequeued first.

4.5.3 Local RRM Tables

Each station maintains a local RRM table which serves as a distributed container for resource allocation information. Table entries correspond to edges $e = (u \rightarrow v)$, similar to the entries stored in NDP tables. As we have seen in the requirements analysis, the depth of topology information that needs to be stored in the RRM table depends on the constraint set, as specified in table 4.2. An RRM table entry comprises the following elements:

- **Transmitter hardware address**. This item equals u's hardware address.

- **Receiver hardware address**. Equals v's hardware address.

- **Color**. The color (resource, channel) assigned to e.

- **State**. Keeps track of an entry's assignment state, an important metric in the soft-decision table merging process, which is going to be outlined below. Each entry might take one of the following states:

  - **Unassigned**. Means that the color field does not contain a valid assignment.

  - **Pre-assignment**. The color field contains a valid assignment. However, the coloring has not been consolidated by negotiation with neighboring stations using the transaction-based voting service. As a consequence, the assignment is locally legal, but might cause a constraint set violation in the big picture, i.e. it is not guaranteed to be globally legal.

  - **Pending transaction**. A pending transaction concerned with this assignment currently exists. The concept of transactions will be described later on.

  - **Affirmed**. The local station has approved a foreign assignment.

  - **Active**. The assignment proposal has been approved and has reached the stage of consensus, i.e. active links are guaranteed not to violate the constraint set even in the global perspective – provided that the network topology does not change.

- **Assigner hardware address**. Identifies the station w that has performed this particular assignment. Because of the global uniqueness property of hardware addresses, this information can be used "as a last resort" to break ties that eventually occur during table merger, i.e. the hardware address serves as the final hard-decision criterion.

- **Assigner degree**. The assigning station's total degree $d_G(w)$, i.e. the count of incoming and outgoing edges incident on w. This is a further soft-decision metric.

- **Sequence**. A sequence counter for this particular connection. Use and maintenance of the sequence counter is described in the course of this section.

When the assignment problem is vertex-based, edges do not have a color assignment: they merely serve as connectivity information base. In this case, additional entries for each vertex within constraint set hop range exist. Here, the transmitter hardware address equals the hardware address of the vertex in question, while the receiver address is set to the predefined broadcast address of "all ones". Color assignments and associated information are then stored in these special-purpose vertex entries.

4.5.4   Soft-decision Table Merging with Hard-decision Fall-back

At different places within execution of TBSD-RRM it becomes necessary to merge two distinct RRM tables into a new one. For example, assume that a station receives an RRM table from one of its neighbors, and wishes to incorporate this information into its own local perspective on the joint assignment task. In this case, table fusion must be performed in such a way that the resulting table does not contain ill-assignments, i.e. assignments that do not obey to specified constraints.

Obviously, we cannot simply append the new table at the end of the original one, because both tables could contain assignments for the same element; even if not, two assignments originating in different tables could be mutually constrained, when both tables are joined together. Therefore, an innovative soft-decision table merging operation with hard-decision fall-back is introduced, which performs table fusion subject to the constraint set. Being a key component of TBSD-RRM, this operation has been designed to support fast convergence of the overall distributed DCA solution. We are going to define a very special union operator that actually carries out table fusion.

Formal Description of RRM Tables

First, let us put the concept of RRM tables into a formal description. Given a constraint set C, an RRM table presents a C-constrained set $\mathcal{A}$ of $N \in \mathbb{N}$ assignment vectors $\mathbf{a}_k$, i.e.

$$\mathcal{A} = \{\mathbf{a}_1, \mathbf{a}_2, \ldots, \mathbf{a}_N\} \tag{4.4}$$

where each assignment vector

$$\mathbf{a} = (e, c, \mathbf{d}) \tag{4.5}$$

holds information about the edge[8] e to be assigned, the color c for this assignment, i.e. $\mathrm{color}\,(e) \leftarrow c$, and a decision vector

$$\mathbf{d} = (s, d_G\,(w)\,, w, q)\,. \tag{4.6}$$

Here, $s \in \{0, 1, 2, 3, 4\}$ denotes the state[9], $d_G\,(w)$ the assigner's total degree, w the assigner's hardware address, and q the sequence number.

Constraint-Aware Union Operator

Let $\mathcal{A}$ be a set of C-constraint assignments on the local graph $G_{\mathcal{A}} = (V_{\mathcal{A}}, E_{\mathcal{A}})$, which is a limited view of the global graph $G = (V, E)$, i.e. $G_{\mathcal{A}} \subseteq G$. Also, let $\mathcal{B}$ denote another set of

---

[8] Remember that vertex assignments are covered by special-purpose edges, where the receiver hardware address matches the broadcast address of "all ones"

[9] State values map to verbose states described earlier, i.e. unassigned $= 0$, pre-assignment $= 1$, etc.

C-constraint assignments on $G_\mathcal{B} = (V_\mathcal{B}, E_\mathcal{B})$, $G_\mathcal{B} \subseteq G$. Then, we define the **constraint-aware union** operation with respect to constraint set C such that

$$\mathcal{U} = \mathcal{A} \bigcup_C \mathcal{B}, \qquad (4.7)$$

yields a set of C-constraint assignments on the graph $G_\mathcal{U} = G_\mathcal{A} \cup G_\mathcal{B}$. Thereby, it is not stated how the union is built, only that $\mathcal{U}$ maintains the property of being C-constraint. For example, an operator implementation that simply discards all assignments, leads to $\mathcal{U} = \emptyset$. This would be a valid (though trivial and useless) C-constraint assignment. In general, there are numerous potential ways of realizing the union.

Progressive Constraint-Aware Union

An algorithm has been designed, which "walks" through the set of candidates $\mathcal{V}$ in a progressive fashion, that is, assignment-by-assignment. However, a candidate assignment

$$\mathbf{v}_k = (e_k, c_k, \mathbf{d}_k) \in \mathcal{V} = \mathcal{A} \cup \mathcal{B}, \qquad (4.8)$$

is picked in a specific order at each step, i.e. the randomly organized candidate set $\mathcal{V}$ is first sorted subject to certain rules to obtain the set of sorted assignments $\tilde{\mathcal{V}}$.

Binary Predicate

We define a binary predicate $\mathrm{less}\,(\mathbf{a}_1, \mathbf{a}_2)$ which compares two assignment vectors $\mathbf{a}_1$, $\mathbf{a}_2$ and decides whether $\mathbf{a}_1$ is "less than" $\mathbf{a}_2$, or not. In this sense, if less equals 1, we say $\mathbf{a}_1 < \mathbf{a}_2$, otherwise, when less equals 0 then $\mathbf{a}_1 \geq \mathbf{a}_2$. The following heuristic rule set is employed to determine whether $\mathbf{a}_1 < \mathbf{a}_2$:

- **Identity Rule.** If $e_1 = e_2 \wedge q_1 \neq q_2$, i.e. when both assignments are concerned with the same edge e but have different sequence numbers, then the decision is solely based on sequence numbers, i.e.

$$\mathrm{less}\,(\mathbf{a}_1, \mathbf{a}_2) = \begin{cases} 1 & \text{for } q_1 < q_2 \\ 0 & \text{for } q_1 > q_2 \end{cases}. \qquad (4.9)$$

  Notice that for $q_1 = q_2$ the common rule applies.

- **Common Rule.** Similar to the previous rule, the vector comparison is broken down into a sequence of scalar vector element comparisons. The order, in which elements are compared, determines the overall comparison result. The following order shall apply:

  1. First compare states s_1, s_2,
  2. then degrees $d_{G_1}\,(w_1)$, $d_{G_2}\,(w_2)$,
  3. then assigner hardware addresses w_1, w_2,
  4. then transmitter hardware addresses u_1, u_2,
  5. then receiver hardware addresses v_1, v_2,
  6. finally sequence numbers q_1, q_2.

  Comparison of two scalar values h_1, h_2 ($h = s, w, u, \ldots$) is done in such a way that less shall return 1 when $h_1 < h_2$, and 0 when $h_1 > h_2$. For $h_1 = h_2$ the next metric is taken into account, and so forth. If all metrics were equal – even after comparing the last ones: q_1, q_2 – then less shall return 0.

Sorted Assignments

Now we make use of the binary predicate less and a suitable sorting algorithm to create $\tilde{\mathcal{V}}$, such that

$$\text{less}\,(\tilde{\mathbf{v}}_i, \tilde{\mathbf{v}}_j) = 1 \qquad \forall\, 1 \leq i, j \leq |\tilde{\mathcal{V}}|, i < j \tag{4.10}$$

holds true. A number of well-known sorting algorithms can be used for this purpose, which are well beyond the scope of this work – refer to [95] for examples.

Building the Union

Finally, the constraint-aware union is determined by iterating through $\tilde{\mathcal{V}}$ **in reverse order**, starting form the last element $\tilde{\mathbf{v}}_{|\tilde{\mathcal{V}}|}$, continuing with $\tilde{\mathbf{v}}_{|\tilde{\mathcal{V}}|-1}$, and so forth, until the first element $\tilde{\mathbf{v}}_1$ has also been processed. In each step, $\tilde{\mathbf{v}}_i$ is included in $\mathcal{U}$ if, and only if...

1. there is no assignment in $\mathcal{U}$ concerned with the same edge e_i, which has a state other than "unassigned" [10], and

2. $\mathcal{U}$ would still remain C-constraint after including $\tilde{\mathbf{v}}_i$. Obviously, this constraint check must be performed on the joint graph $G_\mathcal{U}$.

Otherwise the particular candidate assignment $\tilde{\mathbf{v}}_i$ is discarded, that is, its state is set to "unassigned", which keeps connectivity information, but effectively discards the color assignment. If, for a single edge, two assignments exist (i.e. one from set $\mathcal{A}$, another from set $\mathcal{B}$), the corresponding sequence counter is set to the greater of both.

Discussion

The resulting partially assigned graph is guaranteed to meet constraint-set requirements. Additionally, the common rule enforces that incompatible assignments are cleared out in such a way that assignments with a higher state are kept, while those with lower state are dropped. When states of incompatible assignments are equal, the assignment with highest degree $d_G\,(w)$ will survive, which is a reasonable heuristic, in order to speed up convergence: clearly, assignments that affect only a single, or a few stations are less costly to change than assignments affecting a bulk of stations. However, when degrees are also equal, the assigner's globally unique hardware address is used as a "last resort" to break ties. The identity rule ensures that, for the same connection, the most recent assignment prevails, because "fresh" entries (those with higher sequence numbers q) supersede and replace older ones.

Finalizing Table Fusion

After the constraint-aware union operator has been applied to merge two RRM tables, some edges may have remained unassigned. In order to come up with a new (valid) assignment for these edges, a slightly modified version of UxDMA is run over the partially assigned graph $G_\mathcal{U}$. This yields an assignment set with full coverage over $G_\mathcal{U}$, that is, a set, where all elements (either vertices or edges) will have received a legal color assignment. The slight modification to UxDMA is required to make it accept a partially colored graph, i.e. a graph where some elements are assigned, while others are not. Revising listing 4.1, it is sufficient to drop the following initialization:

---

[10]An element with state = "unassigned" might be safely replaced, as it merely serves as connectivity information

---

for all elements e of type Ψ **do**
  $\text{color}(e) \leftarrow 0$

Running this modified version of UxDMA yields a graph with legal C-constraint assignments for all elements in question, where only previously unassigned elements will have received a new color, while previously assigned elements will have kept their colors.

Maintaining the Sequence Counter

The following procedure shall apply whenever a station assigns a color c to an uncolored edge $e = (u \rightarrow v)$. In this case, the station must...

- set $s = 1$ (i.e. state = "pre-assignment"),

- set $d_G(w)$ to its own total degree,

- set w to its own hardware address,

- and increment the sequence counter by one.

Again, there are some design choices to take regarding which stations shall assign colors to uncolored elements, causing the sequence number to be incremented, etc. For example, it would be possible to let all stations arbitrarily assign colors. However, at present time, stations will only assign colors to their own out-going edges.

4.5.5 Distribution Service

This service provides a means to distribute assignment information within the local neighborhood. For example, when an assignment has reached the "active" state, this information is spread out using the distribution service. This service is able to deliver a copy of the local RRM table (or a subset thereof) to immediate neighbors via MMPDUs. Peer instances incorporate received RRM tables by means of table fusion. However, only those assignments are taken into account, which do not exceed the hop limit for the particular constraint set in use. When changes in activity are detected, these changes are in turn also queued for re-distribution. This mechanism may also be used to deactivate an active link, simply by distributing an assignment with a sequence count higher than that used for activation, and a state other than "active", for example "pre-assignment", "pending transaction" or "affirmed".

4.5.6 Voting Service

As already mentioned before, only activated links are allowed to be used for actual data transmission. A sequence of message exchanges and associated state transitions finally leads to activation of a link. Transactions are a key concept in managing global states in distributed systems. The technology behind the voting service shares some aspects with the concept of distributed transactions, but there are also some differences. In our particular case of distributed DCA, transactions are used to achieve activation of assignments. Let us recall two basic properties of transactions [90], which are also fulfilled in the TBSD-RRM voting service:

- **Atomicity**. A transaction must be "all or nothing", meaning that it either completes successfully, and the effects of all of its operations are recorded in the stations, or it has no effect at all.

- **Consistency**. A transaction takes the system from one consistent state to another consistent state.

To be more specific, a transaction might either succeed, or fail. When it succeeds, it has led to a consistent state transition in local state instances of a global state – here: activation of an assignment. When it fails, it has no effect at all, in the sense that it does not interfere with other concurrent transactions. Some transaction models compensate for node failures only, while others are also capable of tolerating communication errors (packet loss). Some require a common time base (network synchronization), while others can do without. The scheme employed in the context of this work does not require synchronization, and is able to tolerate node failures (exhausted power-sources, software crashes, etc.) and communication errors.

Management Frame Subtypes

The voting service communicates with peer instances by exchanging MMPDUs via the IEEE 802.11 MAC service. Four types of distinct control messages are defined (in addition to the one defined by the distribution service):

Vote Query Initiates a vote query. A station in need of an active link creates a new transaction, enters the "collecting votes" state, and broadcasts a vote query carrying an assignment proposal, thereby inviting its neighbors to cast their votes on this particular assignment. The advanced concept of reliable, yet efficient broadcasts (as introduced in section 3.1) is once again used for this purpose.

Positive Vote A station that receives a vote query either affirms or rejects the assignment proposal. It shall accept the proposal, when it does not conflict with existing assignments or proposals with higher priority. In other words, a positive vote is cast, when the proposal survives a table merge operation, where the two tables involved in table fusion are. . .

- the station's local RRM table, and

- a table only consisting of the assignment proposal

In this case, the station's local RRM table is replaced by the outcome of this table fusion. Other stations that receive the vote packet incorporate positive vote information into their own tables with a state of "pending transaction".

Negative Vote In the opposite case, i.e. when the assignment proposal did not prevail in table fusion due to conflicts with higher-prioritized mutually constraint assignments, a negative vote is cast. The negative vote packet also contains the set of conflicting assignments, so as to deliver the information required for a more successful reattempt.

Positive Announcement When $E_{rr}^1 \in C$, a positive vote must be announced, before the actual vote is cast. This is necessary to avoid a situation, where it cannot be guaranteed that active links always obey to the constraint set. This situation is due to the great separation of managing vertices for this constraint, and is shown in figure 4.9(a): Assume that two concurrent queries (originating in stations one and four) propose the same channel for mutually constraint elements at the same time. The query is queued (in-box), before it is actually transmitted over the air (envelope). Station two receives one's query, while three receives four's query. Both

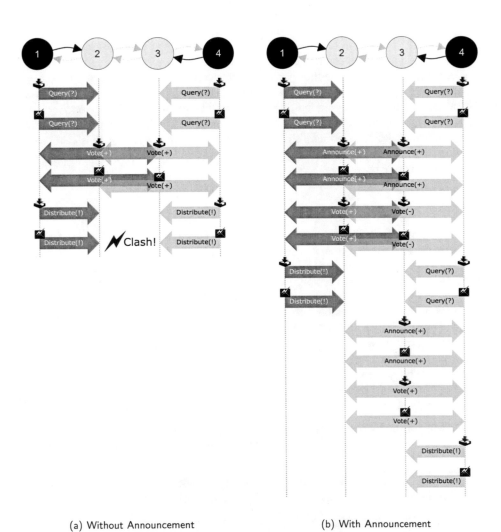

(a) Without Announcement   (b) With Announcement

Figure 4.9: Announcements in Presence of E_{rr}^1 Constraint

accept their respective proposals, because they do not collide with existing assignments. Then, stations one and four activate the assignment, because all due votes have been collected. Bang! There are two mutually constraint assignments, which have reached the "active" state.

One way to prevent this situation is to include the two-away neighbors in the set of dependencies, and let appropriate intermediate one-hop neighbors forward query and response frames. Another solution, which is clearly more scalable, is that of introducing another (additional) type of control frame: positive vote announcements. Now, the situation is different: before a positive vote is cast, an announcement has to be sent first. Station two receives the announcement sent by three, and vice versa. The response packet (positive or negative vote) is queued for transmission, after the announcement has actually been delivered over the air. Similar to positive vote packets, stations that receive the announcement incorporate this information into their own tables, but in this case, the state is set to "affirmed". In the present example, this leads to the assignment being cleared out of station three's RRM table, due to the behavior of operator less. As soon as the announcement has actually been sent, the station casts a positive vote – provided that the assignment has survived until that time; if it has not, a negative vote is cast. It should be stressed that announcements are only required if $E_{rr}^1 \in C$. When $E_{rr}^1 \notin C$ managing vertices are separated by two hops at most.

Local Transaction Tables

Stations maintain a local table for pending transactions. Transactions have limited lifetime only, thus there is no one-to-one correspondence with the RRM table. But at some point in time, there must have existed a transaction for all activated links, because activation is the outcome of a successfully completed transaction. Each transaction table entry is structured in this way:

- **Assignment.** An RRM table entry (or assignment vector), identifying the particular assignment, which is in the scope of this transaction. Transactions may only be created for such assignments, which are also stored in the local RRM table.

- **State.** Each table entry may take one of the following states:

  - **Inactive.** The initial state for any newly created transaction.

  - **Collecting Votes.** A station willing to activate a certain assignment initiates a vote query. This involves creation of a new transaction set to this very state, while votes are collected from neighbors.

  - **Announced Positive Vote.** A positive vote has been announced, but the announcement has not been delivered over the wireless medium yet.

  - **Cast Positive Vote.** A vote query has been received, inviting votes for an assignment proposal. The assignment proposal was compatible with existing assignments, hence, a positive vote has been cast.

  - **Cast Negative Vote.** Similar to the previous case, a vote query has been received. But in this case, the assignment proposal was incompatible with existing assignments. Therefore, a negative vote has been cast, rejecting the proposal.

  - **Active.** After collecting all due votes, the assignment proposal has reached the stage of consensus and is activated. Neighbors are informed about activation via the distribution service.

- **Invited?** Indicates whether the particular station has been invited to vote on the query. Initially, all neighbors are invited to vote, but in a second attempt, for instance, a station is only invited, when its vote was missing previously.

- **Dependencies.** A set of neighbors whose votes are missing for a particular query. Initially all neighbors belong to this set, before individual stations are cleared out of the set when a corresponding vote is received.

- **Votes.** Positive and negative votes are collected in this set. When a vote is received, the originator of the vote is removed from the set of dependencies and inserted into the set of votes, together with an indication whether the vote has been positive or negative. This way, a single station's vote is guaranteed to be counted only once.

- **Reasons.** Negative vote packets usually contain a set of assignments, the negative vote reasons that caused the negative vote. That means, having accepted the vote query would have caused conflicts with these assignments. The station initiating vote query can incorporate these reasons into its own RRM table, thereby extending its own information base. This enables the station to come up with a new assignment proposal – now based on a widened perspective.

- **Lifetime.** Transactions have limited lifetime only, so as to protect the queries from node failures and message loss. When votes are still missing after the lifetime has expired, a retry is started – provided that the attempt limit has not been reached yet.

- **Attempt No.** A station might receive negative votes for transactions in the "collecting votes" state, or due votes may be missing, when transaction lifetime has expired. In this case, another attempt is made to activate the assignment (provided that the attempt limit has not been reached), increasing the attempt count by one. If the set of received votes contains one or more negative votes, the corresponding negative vote reasons are incorporated into the station's local view by table fusion. Two cases may be distinguished at this point:

  1. Table fusion has lead to a revised assignment proposal. In this case, the set of received votes is cleared and the set of dependencies is made equal to the set of all immediate (one-hop) neighbors.

  2. Table fusion did not take place (there were no negative votes), or the previous assignment survived the fusion without modification. Then, the set of received votes is not cleared, effectively keeping positive votes. What is more, the set of dependencies shall only include stations whose votes are missing.

  Finally, another query is invoked.

4.5.7 Priority Queue

The priority queue accounts for properly ordered transmission of RRM control messages. Transaction and distribution service do not request the MAC layer to transmit the control messages immediately, but they push their messages onto the priority queue, which has been introduced earlier (3.2.2). This queue is not organized as a simple FIFO queue. Instead, messages are ordered according to their **priority**, that is, the higher the priority value, the sooner the message is delivered. Priority values for control frames introduced in this chapter are stated in table 4.3.

Message Type	Priority
Vote query (initial attempt)	lowest
Vote query (reattempts)	low
Positive Vote	normal
Negative Vote	normal
Distribution	above normal
Positive Announcement	highest

Table 4.3: Priorities for RRM Control Messages

4.6 Results and Performance Evaluation

In order to evaluate the performance of TBSD-RRM, the protocol has been implemented into the innovative simulator, which is going to be presented in part II. Four representative topologies are used to assess quality of achieved assignments. Considering any particular topology, it can generally be stated: the fewer colors assigned, the better the algorithm.

When comparing the centralized UxDMA with the fully distributed TBSD-RRM, it is obvious that the centralized version is expected to perform better than any distributed solution, due to the advantage of a broader view of the problem (complete topology knowledge). Also, TBSD-RRM has been designed not to require multiple passes over the network to achieve valid assignments.

Obviously, when a distributed algorithm runs several times over the entire network, similar assignment quality can be achieved as in the centralized case. For example, alternate algorithms based on token-passing require several passes over the network [93] – thereby allowing a distributed implementation of the MNF and PMNF heuristics. In a first phase the processing order would be determined, while in the second phase the actual assignments are performed in the order determined in phase one. However, a number of passes is required to determine the processing order, where quality of the ordering achieved in the first phase depends on the number of passes.

It should be stressed that TBSD-RRM does not require multiple passes. Only in case of conflicts, retries are necessary. With each retry, probability of successful activation increases, because of the information gain achieved by piggybacking conflict resolution information into negative votes. Moreover, these retries are of limited scope only, that is, they do not affect distant regions. As a consequence, the assignment results achieved by TBSD-RRM depend on the order of link activation.

Notice that the same heuristics which are available in the centralized case are also available in the distributed case. However, in TBSD-RRM these heuristics only affect the ordering within individual local graphs. In the centralized case, we would expect PMNF to perform best, followed by MNF as second best. The least performance can be expected from the random scheme. Comparing UxDMA with TBSD-RRM, we would also expect that UxDMA outperforms TBSD-RRM due to the information advantage.

In the simulation runs used to obtain the following results, link activations are requested concurrently, probably leading to conflicts and several retry attempts. While it is very unlikely that all users in a network activate their links exactly at the same time, this scenario of simultaneous activation presents the most challenging manifestation of the problem. Recall that management frames are transmitted over the RACH, which is managed by the IEEE 802.11 DCF. As a result, link activation happens in random order, because of DCF's random back-off strategy.

Initially, a state-less hard-decision implementation of the local graph coloring approach had been developed [96]. While this protocol was not modular, and did not offer the benefits of the stateful transaction-based soft-decision RRM protocol presented in this chapter, it had been a

Figure 4.10: Topology "chain10" (Numbers Denote Total Degree)

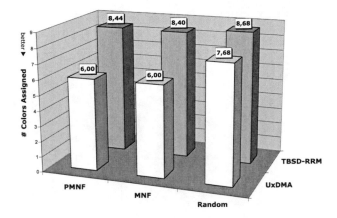

Figure 4.11: Results for Topology "chain10", Averaged over 25 Simulation Runs

promising starting point for the development of TBSD-RRM. A number of simulations had also been conducted, and color histograms had been provided.

4.6.1 Topology "chain10"

In the "chain10" topology, ten stations are arranged in-line to form a chain, as depicted in figure 4.10. Each station can communicate with two neighbors, except for the outer two stations, which have one neighbor each. This results in 18 directional edges, an average total degree of 3.6, and a maximum total degree of 4. This simulation setup is advantageous to study effects of concurrency, due to the ratio of comparably large network diameter and few number of stations. Simulation runs are completed in short time because, both, the total number of stations and the network density are small. On the other hand, there is enough potential for real concurrency and possible conflicts.

A comparison between assignments obtained by the different flavors of UxDMA and TBSD-RRM is provided in figure 4.11, where $C = \left\{ E_{rr}^0, E_{tr}^0, E_{tt}^0, E_{tt}^1, E_{rr}^1, E_{tr}^1, E_{rt}^1 \right\}$. The number of colors assigned are shown[11], averaged over 25 simulation runs. First of all, it can be seen that UxDMA/PMNF and UxDMA/MNF perform equally well (6 colors), while UxDMA/Random results in a 28% increase in the number of colors assigned (7.68 colors on average). This sounds much, but in absolute numbers, less than two colors more have been assigned by the random variant. The distributed TBSD-RRM performs worse in all its variants: it assigns two to three colors more than actually required – a penalty which is caused by concurrent transactions. TBSD-RRM/MNF performs marginally better than TBSD-RRM/PMNF, and slightly better than TBSD-RRM/Random. It is worth mentioning that UxDMA/PMNF is generally superior to UxDMA/MNF, while this relation does not map to TBSD-RRM, because of its incomplete topology knowledge.

---

[11]Notice that in the distributed version the highest color index c_{max} plus one is used as a metric, since it may happen that one or more colors in the range $c_0 \ldots c_{max}$ are not assigned to any link due to concurrency effects

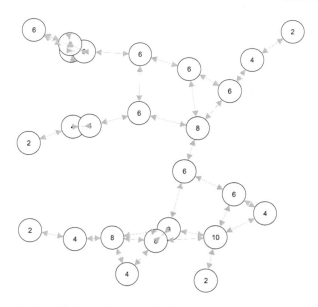

Figure 4.12: Topology "sparse25" (Numbers Denote Total Degree)

4.6.2   Topology "sparse25"

Another relatively sparse topology consisting of 25 stations, "sparse25", is illustrated in figure 4.12. In this scenario, stations have been randomly deployed in a square area, resulting in
68 directional edges. Average and maximum total degrees equal 5.44 and 10, respectively. Like
with the previous simulation series, the full edge constraint set has been specified, as it belongs
to the most challenging problem class.

Again UxDMA/PMNF and UxDMA/MNF perform equally well, but this time the centralized
random variant performs only 3% worse. It is most interesting that the MNF version of the
distributed TBSD-RRM outperforms the centralized UxDMA/Random, and achieves the same
performance as UxDMA – for this particular network. PMNF and random variants of TBSD-RRM
yield results worse by 10% and 18%, respectively. Even though not as good as MNF, the other

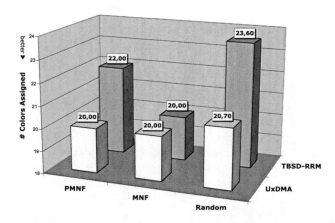

Figure 4.13: Results for Topology "sparse25", Averaged over 10 Simulation Runs

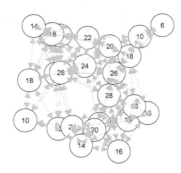

Figure 4.14: Topology "dense25" (Numbers Denote Total Degree)

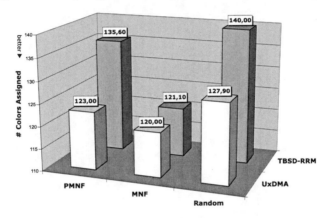

Figure 4.15: Results for Topology "dense25", Averaged over 10 Simulation Runs

two flavors of TBSD-RRM still produce superior assignments than many other distributed DCA algorithms.

4.6.3 Topology "dense25"

Topology "dense25" (figure 4.14) also consists of 25 stations, but now the network is rather dense: this time, 230 inter-station links exist; total degree equals 18.4 on average, and 28 at maximum. Once more, the full set of edge constraints has been specified.

The interesting fact here is that UxDMA/MNF outperforms UxDMA/PMNF – a case, which has occurred only once in the vast number of topologies that have been simulated. Notice that a lot of topologies have been studied, from which only four representative ones have been chosen for discussion in this section. What is even more interesting, is that TBSD-RRM/MNF also outperforms UxDMA/PMNF (not to mention UxDMA's random flavor), while PMNF and random variants of TBSD-RRM are clearly inferior.

4.6.4 Topology "heterogeneous100"

The last topology, called "heterogeneous100", comprises one hundred stations, randomly deployed within a square area, as shown in figure 4.16. Network density is heterogeneous, i.e. some regions are quite densely populated while others are sparse. Expressed in numbers: 508 edges exist,

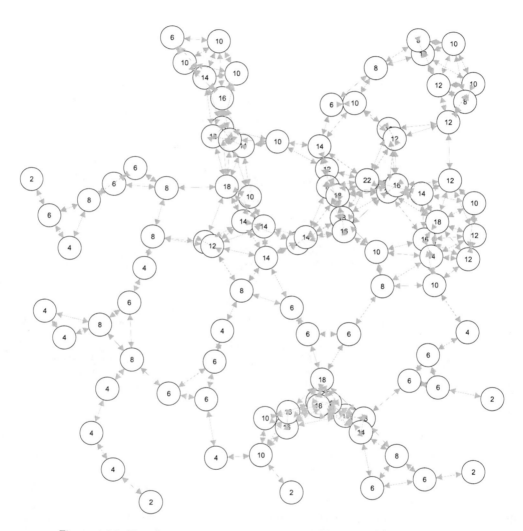

Figure 4.16: Topology "heterogeneous100" (Numbers Denote Total Degree)

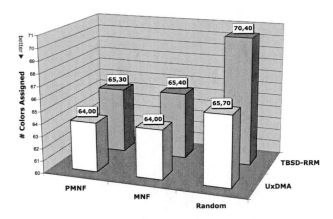

Figure 4.17: Results for Topology "heterogeneous100", Averaged over 10 Simulation Runs

average total degree is 10.16, maximum is 22. In contrast to previous simulation runs, another constraint set (this time describing the TDMA/FDMA link schedule/assignment problem) has been used, i.e. $C = \left\{ E_{rr}^0, E_{tt}^0, E_{tr}^0, E_{tr}^1 \right\}$.

The centralized flavors of UxDMA behave as expected; Random requires about 3% more physical channels than PMNF and MNF. Quality of the colorings produced by TBSD-RRM's PMNF and MNF variants is again close to that of their centralized counterparts. Both variants slightly outperform UxDMA/Random. The worst performance is achieved by the random version of TBSD-RRM, resulting in a 10% waste of radio resources.

4.7 Conclusion

In this chapter a novel modular protocol suite for radio resource management in existing and future wireless networks, in particular multihop ad hoc networks, has been presented. It is worth emphasizing that TBSD-RRM is not only an algorithm, but a ready-to-use protocol, which exchanges management messages over a random access channel, coping with possible node and communication failures. It has been implemented and verified in the simulation environment to be outlined in part II of this work.

What also makes TBSD-RRM different from most other solutions is its ability to operate in an incremental fashion: it is not necessary to run the algorithm (possibly several times) over the entire network, but it is sufficient to activate those links that are actually used, as depicted in figure 4.18 for two bidirectional data flows. Stations that are affected by foreign activations learn which resources are occupied and can effectively reuse this information to come up with valid assignment proposals for their own links. These proposals are very likely to be accepted. Only multiple concurrent activation requests can potentially cause conflicts, which are resolved in repeated activation attempts based on a broadened perspective.

At least three positive implications follow directly from this feature: First of all, links are activated – and hence, ready to be used – rather quickly. Compared to other distributed DCA algorithms, the idea of active and inactive links better meets "real-world" requirements. Moreover, spectrum utilization is further improved thanks to statistical gains: only active links occupy radio resources. Mobility support is inherently provided by the fact, that only local topology information is required to achieve legal assignments. In addition, this new approach facilitates local repair and reassignments in a mobile scenario.

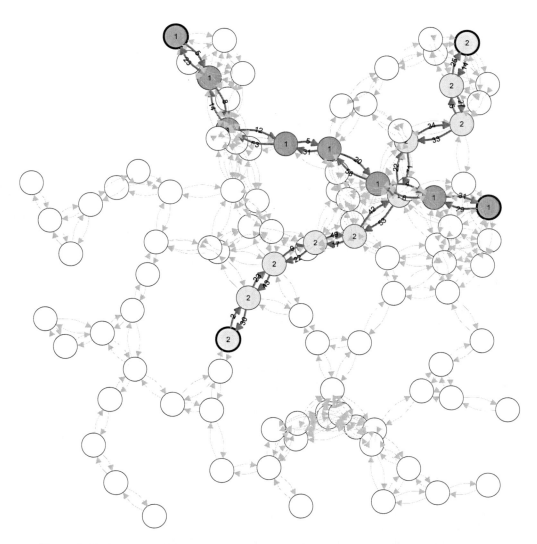

Figure 4.18: Incremental Operation of TBSD-RRM: Two Flows in "heterogeneous100"

The modular architecture facilitates future improvements of the proposed protocol. For example, when a new DCA algorithm becomes available which improves upon UxDMA, it would be easy to integrate into the modular framework. Moreover, modifications to the soft-decision vector and the way how two such vectors are compared (binary predicate) allows changing the assignment strategy while keeping essential features like the transaction-based voting service untouched. Vice versa, when the underlying transport mechanism used to deliver management frames changes dramatically[12], few adaptations to the voting and distribution service would be sufficient to cope with the new situation.

In future work, an optional pre-coloring stage could be added to TBSD-RRM. Such a pre-coloring might be useful for further improvement of spectrum utilization in stationary networks. For example, an algorithm which runs in several distributed passes over network might achieve UxDMA/PMNF-like performance in networks with seldom topology changes. The presented stateful approach is completely orthogonal to such an algorithm. These assignments would be given a low priority state, namely "pre-assignment". In this sense, a pro-active fine tuning of resource allocations becomes a viable option. A further easy to add improvement would be a "blacklist" which avoids usage of channels exposed to extrinsic interference.

---

[12]For example, consider the case when TBSD-RRM was to be used for load-adaptive DCA in existing cellular networks. Then, inter-BS management frames would be exchanged over reliable backbone links

This page intentionally left blank.

CHAPTER 5

Wireless Sensor Networks -
A New Field of Application for Kalman Filtering

This chapter is dedicated to the emerging sector of wireless sensor networks (WSNs). As we will see later on, these share many technological ideas with MANETs, such as the ad hoc and multihop paradigms. On one hand, sensor networks pose very stringent requirements regarding scalability and energy consumption, which are the primary optimization criteria. On the other hand, data rates are very low and mobility is usually none of a concern. Once deployed, a node usually does not move to a different location. Changes in ambient temperature, for instance, may be transmitted in intervals of minutes and hours.

In WSNs, sensor data is usually distributed using a connection-less, multicast-capable transmission service. However, especially in the error-prone wireless environment, packet loss and varying transport delays are common phenomena. In this chapter, a new way of improving reliability of sensor data distribution without compromising network efficiency is presented. Even better, it is shown that, when using this novel paradigm, transmission interval lengths may be increased. Naturally, this results in a gain in network capacity and, as a matter of fact, reduction of overall power-consumption. The proposed solution is applicable to wireless networks in general, and sensor networks in particular. Part of this work has been published in [97].

5.1 Introduction

When having a look at WSNs, usually a large amount of tiny control systems equipped with low-cost radio transceivers form a self-organizing ad hoc network. Being a key technology for new ways of interaction between computers and the physical world around us, WSNs are faced with a rather unique mix of demanding challenges, like

- scalability

- energy-efficiency

- self-configuration

- inexpensive sensor nodes with limited resources

73

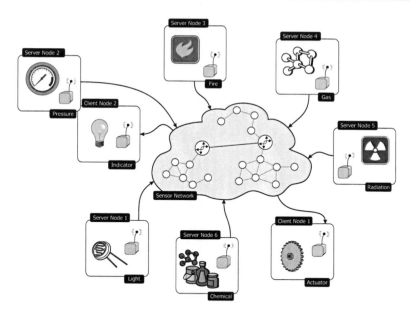

Figure 5.1: Wireless Sensor Network

- data centricity

- ... and many others

Sensor data of different types (temperature, humidity, pressure, lighting condition, battery level, vehicular movement, soil makeup, noise levels, ...), originating in different locations, must be distributed among interested groups within the collective, as illustrated in figure 5.1. This information is required by distributed control algorithms, which often run periodically. Based on this sensor information, output to actuators is calculated. Imagine that actuator output is required every 10 ms, for instance. Then this would mean that sensor input is also required every 10 ms. However, it might not be feasible – and for sure not very efficient[1] – to transmit the sensor data with the required periodicity. In addition, due to the common switching and multiple access strategies employed in WSNs, transport delays show some variation. At times, even complete packet loss may occur, and retransmission is hardly an option in real-time distributed multipoint-to-multipoint systems.

An efficient way to distribute such information is to use the wireless medium's native broadcast feature. Sadly, broadcasts – and multicasts in general – are somewhat unreliable compared to unicasts. Let us consider one of the most popular protocols for distributed random access to a shared wireless channel, IEEE 802.11 DCF, as an example. Then, we can identify the following major reasons for the lack of reliability:

- **Acknowledgement**. Because of the well-known ACK-explosion problem, multicast frames are not protected by immediate positive ACKs, while directed frames are.

- **Fragmentation**. In order to reduce loss probability, large unicast frames are split into several smaller portions (fragments). When a fragment happens to be lost, only the short fragment needs retransmission. Multicast frames cannot be fragmented.

---

[1]with respect to power-consumption, network traffic load, etc.

- **Virtual Carrier Sense**. A priori medium reservation under RTS/CTS regime is not available in multicast operation, resulting in two negative implications. First, packet loss probability is increased, and second, network throughput is degraded in presence of hidden terminals – refer to section 1.5.2 for details.

The source of all of these problems is the point-to-multipoint situation, which also affects network and transport layers. For obvious reasons, TCP is not suitable for multicast. While alternate transmission protocols are in the focus of current research, it is common practice to use UDP instead. Because of its simple nature, UDP is quite an efficient protocol, when packet loss and network congestion are no issues. But for the same reason it is also an unreliable protocol: there is no way for the transmitter to know whether any, some, or all of the addressed stations have successfully received a message.

Although UDP on top of IEEE 802.11 is not a likely combination for wireless sensor networks, some of the identified problems are inherent to many unacknowledged, connection-less, multicast capable transport services and carrier-sense based multiple access protocols. The solution presented in this chapter is generic in nature, and does not rely on UDP. What is really required, is a way to transmit data messages with sensor value samples from the source nodes to a sink. UDP has been selected as one possible example, because of its widespread use.

The new solution is based on the combination of KALMAN filter and TAYLOR approximation techniques, continuously estimating relevant sensor values at the sink. Therefore, a brief overview of KALMAN filter theory is given in section 5.2, and its application to wireless sensor networks is presented in section 5.3. Based on the theoretical foundation, practical implementations are outlined in section 5.4, including a generic variant suited for design and simulation stages, and an example of a highly optimized version specifically tailored to the requirements of tiny sensor nodes. A framework in terms of a protocol specification is given in section 5.5, and compatibility of the proposed solution with existing WSN approaches is discussed in section 5.6. In section 5.7 a further new and promising approach is presented: the application of KALMAN filter methodology for data fusion in multihop WSNs. Numerical results for a real-life sensor signal are presented in section 5.8, before the chapter is finally concluded in section 5.9.

5.2 Kalman Filter

5.2.1 System Model

Considering the state space description,

$$\mathbf{x}\left(k+1\right) = \mathbf{A}\mathbf{x}\left(k\right) + \mathbf{u}\left(k\right) \tag{5.1}$$

$$\mathbf{y}\left(k\right) = \mathbf{C}\mathbf{x}\left(k\right) + \mathbf{w}\left(k\right) \tag{5.2}$$

where $\mathbf{u}$ denotes system noise and $\mathbf{w}$ denotes observation noise, respectively, the KALMAN filter [98–101] is basically responsible for estimating the state vector $\mathbf{x}$ given only the observations $\mathbf{y}$. The filtering operation can be split into two distinct parts: **prediction** (or estimation) and **update**.

5.2.2 Prediction

During a prediction step, a new value for the state vector $\mathbf{x}$ at time k is estimated based on the value of $\hat{\mathbf{x}}$ at time $k-1$ (5.3), leading to a new estimate for $\mathbf{y}$ at time k (5.5). Furthermore, the error covariance matrix $\mathbf{P}$ is adjusted accordingly (5.4),

$$\hat{\mathbf{x}}\,(k|k-1) = \mathbf{A}\,\hat{\mathbf{x}}\,(k-1|k-1) \tag{5.3}$$

$$\mathbf{P}\,(k|k-1) = \mathbf{A}\,\mathbf{P}\,(k-1|k-1)\,\mathbf{A}' + \mathbf{Q} \tag{5.4}$$

$$\hat{\mathbf{y}}\,(k|k-1) = \mathbf{C}\hat{\mathbf{x}}\,(k|k-1) \tag{5.5}$$

where the superscript prime denotes transposition of a matrix. Matrix $\mathbf{P}$ includes information from two other covariance matrices, namely the system noise covariance matrix $\mathbf{Q}$ and the observation noise covariance matrix $\mathbf{R}$, and thus gives a measure of the current **estimation quality**. In general, these noise covariance matrices may be expressed as

$$\mathbf{Q} = \mathrm{E}\,\{\mathbf{U}\,(k)\,\mathbf{U}'\,(k)\} \tag{5.6}$$

$$\mathbf{R} = \mathrm{E}\,\{\mathbf{W}\,(k)\,\mathbf{W}'\,(k)\} \tag{5.7}$$

where $\mathbf{U}$ and $\mathbf{W}$ are stochastic processes describing the two kinds of additive noise mentioned above (5.1, 5.2).

5.2.3   Update

When a fresh update (or observation) $\mathbf{y}$ is available, this update is used to correct the estimated values of $\hat{\mathbf{x}}$ (5.9) and, as a result, of $\hat{\mathbf{y}}$. The KALMAN gain matrix $\mathbf{K}$ is adapted (5.8) and the error covariance matrix $\mathbf{P}$ is recalculated as well (5.10),

$$\mathbf{K} = \mathbf{P}\,(k|k-1)\,\mathbf{C}'(\mathbf{C}\mathbf{P}\,(k|k-1)\,\mathbf{C}' + \mathbf{R})^{-1} \tag{5.8}$$

$$\hat{\mathbf{x}}\,(k|k) = \hat{\mathbf{x}}\,(k|k-1) + \mathbf{K}\,(\mathbf{y}\,(k) - \mathbf{C}\hat{\mathbf{x}}\,(k|k-1)) \tag{5.9}$$

$$\mathbf{P}\,(k|k) = \mathbf{P}\,(k|k-1) - \mathbf{K}\,\mathbf{C}\,\mathbf{P}\,(k|k-1)\ . \tag{5.10}$$

5.2.4   Asynchronous Enhancements

The standard KALMAN filter known from motion tracking applications relies on synchronously and equidistantly available measurement updates. Classical applications, e.g. radar systems used in air traffic control, comply with these requirements. In many other cases, though, the time instants when, (a) input to the filter is available, and (b) output from the filter is desired, may diverge. Obviously, the classical KALMAN filter approach would fail under such conditions. Therefore, enhancements to the standard KALMAN filter have been developed in the context of this work. They make it accept sporadic, infrequent updates, while still producing periodic output. Since input and output sections are decoupled, the resulting structure is called: asynchronous KALMAN filter. In fact, the asynchronously operated KALMAN filter is such a generic tool that it has also been successfully applied to areas other than wireless sensor networks [102, 103].

In order to be capable of handling time varying transport delays, messages containing sensor values are marked with time stamps at the message generation site (the server site). Hence, update messages consist of a sensor value observation $\mathbf{y}\,(k_m)$ and its corresponding time stamp

k_m. For now, let us assume that all processes running on the various sensor nodes share the same notion of time, meaning that they share a common time base. Later it shall be argued that this requirement can be relaxed at the price of an uncompensated delay.

At the client site, messages are received and queued based on their time stamps. At the next synchronous clock instant a message is de-queued and ready to be applied. At this point, we can distinguish three cases when comparing the message's time stamp k_m with the client process's local time k_c:

1. The update becomes valid in the **future** $(k_m > k_c)$. In this case the update is deferred until the local time reaches the message's time. Then this update belongs to the next case. Note that this case has only been introduced for theoretical comprehensiveness, though it may be useful for a few special applications.

2. The update is valid **right now** $(k_m = k_c)$. Then it is applied immediately. This corresponds to the case, when network transmission delay is in the order of time stamp granularity.

3. The update had been valid in the **past** $(k_m < k_c)$. Then the update is applied to a previously saved filter state and $M = k_c - k_m$ predictions are performed in a row.

This way, the filter is capable of handling past, present and future updates. Most of the time, the filter will be processing updates that belong to a past clock instant, caused by some transport delay between server and client. This transport delay can be compensated with a number of prediction steps in a row at the cost of an increasing estimation error ϵ.

$$\epsilon = |\mathbf{y}(k) - \hat{\mathbf{y}}(k|k - M)| \tag{5.11}$$

The overall filter structure is shown in figure 5.2.

5.3 Application to Sensor Values and Wireless Networks

In collision-constrained wireless networks, sensor samples may arrive at random time instants due to transport, network and link layer effects. Sometimes, even complete message loss may occur. Nonetheless, the current sensor value may be desired more frequently, than samples are available. Therefore, predictions of the sensor values in question are performed periodically (synchronously) with the desired output rate.

5.3.1 Generic Deterministic Modelling

When using a KALMAN filter, two deterministic matrices, the system transition matrix $\mathbf{A}$ and the observation matrix $\mathbf{C}$, need to be set up properly. In this section, a generic solution based on a TAYLOR approximation of the desired sensor signal is provided. Thanks to the TAYLOR approximation's generic nature, the solution is applicable to a vast number of entities, for instance, temperatures, pressures, etc. Let $p(t)$ denote the sensor signal in question, then its value at time $t + \Delta t$ is approximated by

$$p(t + \Delta t) = p(t) + \Delta t \frac{\mathrm{d}p(t)}{\mathrm{d}t} + \frac{(\Delta t)^2}{2} \frac{\mathrm{d}^2 p(t)}{\mathrm{d}t^2} + \cdots . \tag{5.12}$$

Now, assume a discrete-time signal with sampling period T. This leads to

$$p(kT + T) = \sum_{n=0}^{N-1} \frac{T^n}{n!} \cdot p^{(n)}(kT) . \tag{5.13}$$

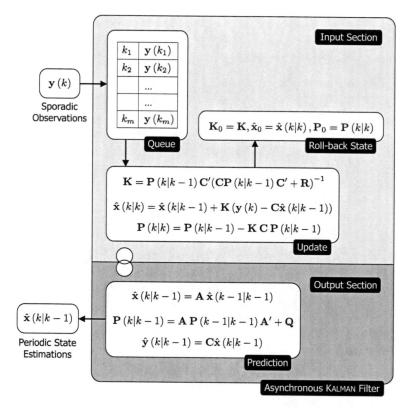

Figure 5.2: Asynchronous KALMAN Filter (Structural View)

Matrix $\mathbf{A}$ expresses the transition from one clock instant to the following; therefore, equation (5.13) yields the structure and contents of $\mathbf{A}$. For practical reasons, the TAYLOR approximation is aborted at some point, for instance after the second derivative term. In this case we would identify this setup as a second order model, i.e. $n = 2$. Dimensions of state vector $\mathbf{x}$ and system transition matrix $\mathbf{A}$ depend on n. While the state vector is given by

$$\mathbf{x} = \left(p, p^{(1)}, p^{(2)}, \ldots, p^{(n)}\right)' , \tag{5.14}$$

the system transition matrix would look like

$$\mathbf{A} = \begin{pmatrix} 1 & T & \frac{T^2}{2} & \cdots & \frac{T^n}{n!} \\ 0 & 1 & T & \cdots & \frac{T^{n-1}}{(n-1)!} \\ \vdots & & & & \\ 0 & 0 & 0 & \cdots & 1 \end{pmatrix} . \tag{5.15}$$

Finally, the observation matrix $\mathbf{C}$ is responsible for gathering the filter's output signal $\mathbf{y}$ by collecting the relevant information from the state vector $\mathbf{x}$. For instance, if we were only interested in the sensor value itself then $\mathbf{C}$ could be expressed as

$$\mathbf{C} = (1, 0, 0, \ldots, 0) , \tag{5.16}$$

where its dimension equals $1 \times (n+1)$. However, if we were also interested in the first derivative, for instance, such information would be very easy to determine, since it is already given in the state vector. The information is then available to adaptive control algorithms, which can modify some of their internal coefficients accordingly. By the way, this model easily extends to multidimensional sensor input, which may be beneficial when correlational dependencies exist across dimensions. Interested readers should refer to [102] for a 3D example.

5.3.2 Generic Stochastic Modelling

In addition to the deterministic matrices $\mathbf{A}$ and $\mathbf{C}$ mentioned in the previous section, two other stochastic matrices have to be designed: The system noise covariance matrix $\mathbf{Q}$ and the observation noise covariance matrix $\mathbf{R}$, respectively. $\mathbf{Q}$ is of the same dimension as $\mathbf{A}$, while, for one-dimensional sensor data, $\mathbf{R}$ is scalar. In order to have these matrices set up correctly, one has to understand the source of the noise they model.

System noise is caused by the fact that, for practical applications, the TAYLOR approximation (5.13) is aborted at some point, e.g. for $n = 2$, derivatives $p^{(3)}$, $p^{(4)}$ and so forth are simply discarded (approximation error). The model implies that the highest order derivative included (in the previous example $p^{(2)}$) remains constant for the duration T. Observation noise may be caused by quantization errors, systematic errors associated with the method of measurement, etc.

Again, let us consider the second order case. System noise is introduced by the fact that the highest order derivative, i.e. $p^{(2)}$ is faulty when the sensor signal's second order derivative (acceleration) changes between two clock instants. We take this into account in the following way. First, we determine the covariance σ^2 for some typical sensor characteristic. Second, we set the corresponding matrix element Q_{nn} to the value of σ^2. Note that the covariance is set only in the highest order term, since it will automatically propagate by means of the system transition matrix $\mathbf{A}$ to all lower order terms.

$$\mathbf{Q} = \begin{pmatrix} 0 & 0 & 0 & \cdots & 0 \\ 0 & 0 & 0 & \cdots & 0 \\ \vdots & & & & \\ 0 & 0 & 0 & \cdots & \sigma^2 \end{pmatrix} \tag{5.17}$$

Observation noise is modelled by its covariance τ^2:

$$\mathbf{R} = \left(\tau^2 \right) \tag{5.18}$$

The last matrix to mention in this context is the error covariance matrix $\mathbf{P}$, which has the same dimension as $\mathbf{Q}$ and $\mathbf{A}$. Since this matrix is adapted automatically at every prediction and update instant, only its initial value $\mathbf{P}(0)$ must be set. We know the sensor's initial value, so we will not make any error there. But because we don't know a second value yet, all derivatives will most probably be faulty in the first place (except for the trivial case, when the value does not change at all), with the highest order derivative being the most faulty. Summarizing these thoughts, we get

$$\mathbf{P} = \begin{pmatrix} 0 & 0 & 0 & \cdots & 0 \\ 0 & a_1 & 0 & \cdots & 0 \\ \vdots & & & & \\ 0 & 0 & 0 & \cdots & a_n \end{pmatrix}, \tag{5.19}$$

where a_i are some constants with $a_{i+1} \geq a_i$.

5.4   Practical Filter Implementations

In this section, two practical implementations for the KALMAN filter are provided. One has been developed with emphasis on general applicability; the other is specifically tailored to tiny devices with scarce processing and memory resources.

5.4.1   Generic Implementation

```
CMatrix CKalmanFilter::Predict()
{
    m_vectorX = m_matrixA * m_vectorX;
    m_matrixP = m_matrixA * m_matrixP * m_matrixA.Transpose() + m_matrixQ;

    return m_matrixC * m_vectorX;
}

void CKalmanFilter::Update(const CMatrix &vectorY)
{
    m_matrixK = m_matrixP * m_matrixC.Transpose() *
        (m_matrixC * m_matrixP * m_matrixC.Transpose() + m_matrixR).Invert();

    m_vectorX = m_vectorX + m_matrixK * (vectorY - m_matrixC * m_vectorX);
    m_matrixP = m_matrixP - m_matrixK * m_matrixC * m_matrixP;
}
```

Listing 5.1: Generic C++ Implementation of the KALMAN Filter

Efficient, optimized solutions have one drawback: they always rely on certain assumptions and constraints on matrix dimensions, symmetry, etc. Therefore, a generic implementation, which is more flexible but less efficient, has been implemented. It is perfectly suited for simulations, and has been implemented as a set of C++ classes. First, a class CMatrix is defined, which encapsulates generic matrix operations, such as addition, multiplication, inversion, transposition, etc. on arbitrarily sized matrices. Then, the class CKalmanFilter defines the filtering process in terms of concatenations of these operations, as outlined in listing 5.1. This yields maximum flexibility (ability to use different TAYLOR approximation orders, arbitrarily structured matrices $\mathbf{A}, \mathbf{C}, \mathbf{Q}, \mathbf{R}$, etc.) and readability of the code at the expense of increased memory and processing requirements.

5.4.2 Efficient Implementation

Different levels of optimization are possible reducing the required computational resources dramatically. These optimizations can be achieved by assuming fixed matrix structures, which are often symmetric and sparse. Furthermore, when performing calculations cleverly, a number of computational results may be reused, while matrix calculation progresses. Using such a progressive technique allows processing of numerous sensor values even with tiny devices offering not much computational capacity.

The results discussed in section 5.8 are based on a simple, one-dimensional, first order model. For this specific case, efficient algorithms are provided to perform the computations associated with KALMAN filter equations. Based on these algorithms, computational and memory requirements in practical applications are evaluated.

Let us consider the equations associated with the prediction step, i.e. equations (5.3, 5.4, 5.5). As already mentioned before, this step is performed quite often, or more precisely, with a frequency equivalent to the desired output rate. When we assume the following,

$$\mathbf{x} = \begin{pmatrix} p \\ \dot{p} \end{pmatrix}, \mathbf{A} = \begin{pmatrix} 1 & T \\ 0 & 1 \end{pmatrix}, \mathbf{C} = \begin{pmatrix} 1 & 0 \end{pmatrix}, \mathbf{P}(0) = \begin{pmatrix} a & b \\ b & c \end{pmatrix},$$
$$\mathbf{Q} = \begin{pmatrix} 0 & 0 \\ 0 & \sigma^2 \end{pmatrix}, \mathbf{R} = \begin{pmatrix} \tau^2 \end{pmatrix}, \tag{5.20}$$

then equation (5.4) manifests as:

$$P_{11}(k+1) = P_{11}(k) + TP_{12}(k) + TP_{21}(k) + T^2 P_{22}(k) \tag{5.21a}$$
$$P_{12}(k+1) = P_{11}(k) + TP_{22}(k) \tag{5.21b}$$
$$P_{21}(k+1) = P_{21}(k) + TP_{22}(k) \tag{5.21c}$$
$$P_{22}(k+1) = P_{22}(k) + \sigma^2 \tag{5.21d}$$

Because of $P_{12}(0) = b = P_{21}(0)$ and (5.21b, 5.21c), it is evident that

$$P_{12}(k) = P_{21}(k), \forall k \geq 0. \tag{5.22}$$

By the way, the same is true for the update step, eliminating the need to calculate and store P_{21} at all. Using a clever interleaving of operations, only three multiply-accumulates (MACs) and one accumulation is required to perform all the calculations stated in equations 5.21. The next step is to update the state vector according to (5.3), which in our specific case simply leads to

$$x_1(k+1) = x_1(k) + Tx_2(k). \tag{5.23}$$

Finally, the output signal needs to be calculated according to (5.5). However, in our case, the output signal is identical to x_1.

In listing 5.2, an efficient C++ implementation of prediction and update steps is given, based on the assumptions made in (5.20). Notice that a plain C or Assembler implementation is readily derived from the C++ representation for small devices where a C++ compiler is not available. The techniques used to obtain the update step's implementation are comparable to those used for the prediction step, hence they are not going to be discussed here in detail.

```
float COptimizedKalmanFilter::Predict()
{
    // P = A * P * A' + Q;
    m_fltP11 += m_fltT * m_fltP12;
    m_fltP12 += m_fltT * m_fltP22;
    m_fltP11 += m_fltT * m_fltP12;
    m_fltP22 += m_fltQ22;

    // x = A * x and
    // y = C * x
    return m_fltX1 += m_fltT * m_fltX2;
}

void COptimizedKalmanFilter::Update(const float fltY)
{
    // Precalculate some intermediate results
    register const float fltM11 = m_fltP11 + m_fltR;
    register const float fltDelta = fltY - m_fltX1;

    // K = P * C' * (C * P * C' + R)^-1
    m_fltK1 = m_fltP11 / fltM11;
    m_fltK2 = m_fltP12 / fltM11;

    // x = x + K * (y - C * x)
    m_fltX1 += m_fltK1 * fltDelta;
    m_fltX2 += m_fltK2 * fltDelta;

    // P = P - K * C * P
    m_fltP22 -= m_fltK2 * m_fltP12;
    m_fltP12 -= m_fltK2 * m_fltP11;
    m_fltP11 *= 1 - m_fltK1;
}
```

Listing 5.2: Efficient C++ Implementation of the First Order Filter

5.4.3   Considerations on Computational Costs

It is worth mentioning that the KALMAN filter is not very costly with respect to computational load when implemented efficiently. Having a look at listing 5.2, the prediction takes five additions, and four multiplications. The update is somewhat costlier: Three additions, four subtractions, five multiplications, and two divisions are required. Note that computations associated with the update are only performed in response to an incoming update sample – an event that occurs sporadically compared to the periodic predictions. Memory requirements are as follows: Seven 32-bit floating point variables are required, and three floating point constants, for a total of 40 bytes. Memory requirements may be further reduced by hard-coding of the constants, resulting in a memory requirement of 28 bytes per filter.

Asynchronous extensions add a slight overhead, which can be kept very small by limiting the queue size to a single sample, i.e. eliminating the queue. A single sample queue does not degrade performance if predictions are performed more often than updates, samples arrive with a delay, and update messages are strictly ordered. These conditions are easily met in WSNs.

Moreover, the filtering operation is valuable from a signal processing point of view. In particular, adaptation of R and Q provides a way of sensor noise reduction. In other words, the computations performed during KALMAN filtering allow to trade-off traffic load reduction against reliability – while at the same time signal quality is improved.

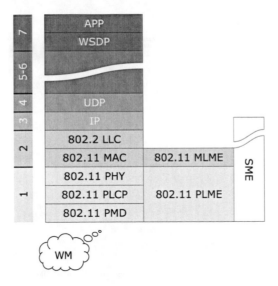

Figure 5.3: Example Protocol Stack for WSDP – Also Used in Simulations

5.5 Wireless Sensor Data Protocol

Based on the theoretical foundation mentioned in the previous sections, a new protocol for transmission of wireless sensor data has been developed: WSDP. Located between application and UDP (or any other datagram service), this middleware component accepts sensor data from applications and utilizes a UDP port for delivery of WSDP protocol data units (WSDP-PDUs), as depicted in figure 5.3. Again, UDP is just an example, almost any datagram service is suitable on top of almost any kind of network – WSDP is an application layer protocol.

The proposed architecture allows a single sensor data server to distribute its information to many clients. And vice versa, a single client may subscribe to a number of data sources. Also, any station may be client and server at the same time, or decide to be either one.

5.5.1 Sending Sensor Data

Server applications willing to transmit sensor data issue a WSDP-DATA.request. This service primitive accepts a time stamp k_m, a sensor value $\mathbf{y}(k_m)$, a unique sensor identifier (SID), and addressing information (destination IP address). The IP address may be a unicast, multicast or broadcast address, while the SID unambiguously identifies the sensor. The request service primitive will construct a datagram and instruct UDP to deliver the message.

5.5.2 Subscribing to Sensor Data

Clients subscribe to sensor readings of a specific sensor simply by invoking a so-called WSDP-SUBSCRIBE.request. In addition to a SID, values for the desired approximation order and output period T are specified, too. From this time on, the client will be notified about the sensor's value through periodic WSDP-DATA.indications, until the subscription is cancelled by a WSDP-UNSUBSCRIBE.request.

Transparent to applications, the client runs an asynchronous KALMAN filter tracking the value of each sensor it subscribes to. Whenever a WSDP-PDU carrying sensor data is received, this data is passed to the filter as an update $y(k_m)$ – provided the update message has not been corrupted due to transmission errors. Transmission errors are easily identified using UDP's CRC checksum. Analogously, when the next WSDP-DATA.indication is scheduled, a KALMAN prediction step is performed and the resulting output $\hat{y}(k_c|k_c - 1)$ is passed to the client. Clients have access to all filter matrices, so they are free to use different tracking models (matrix A), different noise parameters (matrices R, Q), etc.

5.5.3   Retrieving Sensor Data

Clients will be informed by means of a WSDP-DATA.indication about the current state of a sensor value. Previously, they must have subscribed to one or more SIDs. As part of the indication, SID, time stamp k_c, and sensor value estimation $\hat{y}(k_c|k_c - 1)$ are delivered to the client application. In essence, this value is the KALMAN filter's output.

5.5.4   Data Reduction at the Source

Depending on application and available processing power in each node, an additional server side KALMAN filter may be an option. One can think of many ways of limiting the data rate at the source. One way would be to mimic the same KALMAN filter settings, which are also used at the clients. This yields a predictor/corrector setup, in which a threshold ϵ_c on the tolerable estimation error ϵ may be defined. Then, updates would be sent only if $\epsilon \geq \epsilon_c$, for instance. In other words: When sensor values change quickly then updates are sent more frequently, while only few updates are required otherwise.

Another straight-forward way to implement data reduction at the source would be a uniform sub-sampling of sensor data streams. Obviously this solution lacks the adaptivity of the previously described solution. However, it is easy to implement and offers reduced computational cost in comparison to a second KALMAN filter. In addition, uniform sub-sampling helps to increase robustness against erroneous bursts.

The decision whether to use an adaptive scheme or not, should be based on the kind of sensor network used: When observing uncorrelated phenomena, an adaptive scheme may improve bandwidth utilization (statistical multiplexing). On the contrary, when a lot of co-located sensor nodes observe the same phenomenon, an adaptive scheme may increase loss probability due to frequent update packet collisions.

5.6   Compatibility Issues

5.6.1   Synchronization

It was previously stated that nodes are supposed to share a common time base, which requires some kind of synchronization mechanism. Though, time synchronization in large-scale multi-hop ad hoc networks seems to be a rather complicated task. Fortunately, the asynchronous KALMAN filter can be operated without synchronization at all. This comes at the expense of an uncompensated transmission delay. If a common time base is available, the KALMAN filter will automatically compensate for varying transmission delays using the time stamp information provided in the update messages. In contrast, if source and sink nodes are not synchronized, the transmission delay caused by lower layers cannot be compensated anyway.

In this case, the receiver should use the time stamps k_m only for rejection of duplicate frames and detection of mis-sequenced packets. Then, it simply applies the updates **as soon**

as they arrive, effectively bypassing the queue. The filter will output a delayed version of the original sensor data, which is even smoother, compared to the synchronized case with delay compensation [102]. Furthermore, missing synchronization may lead to signal distortions, when jitter (delay uncertainty) is not negligible. Anyhow, these unavoidable distortions are not related to the filtering process, since they would also occur without any filter at all.

5.6.2 Data-Centric Approaches

Although WSDP has been expressed in terms of node-centric UDP, from the discussions in sections 5.2, 5.3 and 5.4 it should be clear that the asynchronous KALMAN filter may be combined equally well with data-centric approaches [104]. From the application layer point of view, such approaches primarily affect addressing issues. For instance, in lieu of IP addresses and SIDs, attribute names and sensor locations may be used. To take it further: Asynchronous KALMAN filter methodology is easily inserted into almost any data flow.

5.7 Data Fusion

Minimizing network traffic – and consequently overall energy consumption, scalability, etc. – is a major concern in WSNs. Advanced techniques have been introduced, including data aggregation (or data fusion), in intermediate nodes. For the first time, an asynchronously operated KALMAN filter is promoted as the ideal, and most natural tool for data aggregation along intermediate nodes.

Keeping in mind that computational costs are generally negligible compared to transmission costs in wireless sensor networks, data fusion, or data aggregation takes full advantage of this notion. Here, instead of simply forwarding sensor value streams originating in different nodes towards interested sites, various sensor-readings from different sources characterizing the same physical phenomenon are combined to form an aggregated signal. Obviously this implies a data centric approach, in which all operations are performed on named data, i.e. (attribute, value) pairs. The results presented in this section are generic in nature, and hence the same ideas are also immediately applicable to sensor network frameworks and communication paradigms other than those presented in the previous sections of this chapter – in particular to directed diffusion [105].

In the previous sections, two promising applications of both, the synchronous and asynchronous KALMAN filter, have been introduced. Using an asynchronous filter at the sink increases the quality of the received sensor data, mitigates the disadvantage of nonuniform packet receptions significantly and allows reducing the update rate dramatically. Furthermore, using a classic (synchronous) filter at the source offers an interesting possibility to decrease the network traffic at a slight expense of "node cost". This is done by letting the filter select only those observations that are required to reconstruct the sensor signal at the sink within a tolerable estimation error ϵ_c.

Another, new application for the asynchronous filter emerges, when we consider intermediate stations along the data flow from source to sink. Such a typical sensor network topology is illustrated in figure 5.4. Here, multiple sources (white) observing the same phenomenon (cloud) disseminate their sensor value streams towards a sink (black) over several hops, utilizing intermediate stations (gray) as relays. Wireless links are denoted by dashed lines, while the data flow is marked by solid lines with arrows indicating its direction.

When a phenomenon's attribute changes it is likely that many nodes in the vicinity increase the frequency of sensor sample transmissions – assuming an adaptive algorithm for update selection at the source. Hence, intermediate nodes will receive multiple samples from different sensor

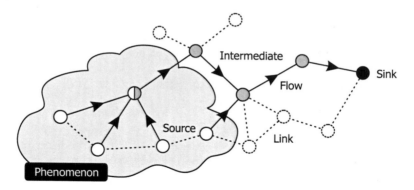

Figure 5.4: Example Topology for Data Fusion in WSNs

nodes characterizing the same physical phenomenon. Plain forwarding of such sensor readings towards the sink is for sure not very efficient with respect to power-consumption, network traffic load, etc. Instead, the received samples are input to an asynchronously operated KALMAN filter (according to their time stamp k_m). Doing so, offers several advantages:

First of all the average sample rate may be noticeably reduced. As already mentioned in the context of filtering at the sink, the update rate may be dramatically reduced thanks to KALMAN prediction. Having a look at intermediate nodes, the situation looks even better. As multiple observation streams are received, the individual streams may be transmitted at even lower sample rates – provided that there is some time diversity between sample instants stemming from different sources. Obviously, the KALMAN filter shows best performance when samples from different input streams are uncorrelated – this maximizes information gain when source samples are combined at the filter's input.

Fortunately, this is a very common and natural situation in wireless sensor networks where nodes are randomly deployed (leading to space diversity), and sampling instants are unsynchronized. In fact, it would be rather difficult to synchronize sample instants amongst different nodes. Even if we assume that source nodes use an adaptive algorithm to distinguish important from less important observations (and thus transmission times are synchronized by the phenomenon itself), extending the adaptive algorithm with a simple randomizer approach again yields comparable performance.

Instead of simply forwarding all incoming source updates, intermediate stations should run an adaptive algorithm[2] to classify those samples (from the set of sensor value estimations) that are worth transmitting. When no such algorithm is available, or computational resources are scarce, uniform sub-sampling may also be applied in intermediate stations.

5.8   Results

5.8.1   Simulation Setup

Simulations have been carried out using the innovative simulator for wireless networks, which is going to be described in part II. Designed especially for wireless ad hoc networks, this simulator perfectly matches the requirements inherent to the concepts outlined in this chapter.

WSDP has been implemented into the simulator and several simulations have been conducted

---

[2]One such algorithm has been presented in section 5.5.4

in order to evaluate different setups. The exact simulation script used to evaluate the filter performance is printed in listing 7.6, belonging to chapter 7, which is concerned with typical simulation flows. Data used in simulations is not artificial. In fact, it comes from a magnetic tracking sensor at a sample rate of approximately 180 Hz. The entire set contains 2873 samples, corresponding to nearly 16 seconds of recorded sensor data. Besides the particular signal presented here, many other signals have been studied, too – always yielding similar performance.

5.8.2 Discussion

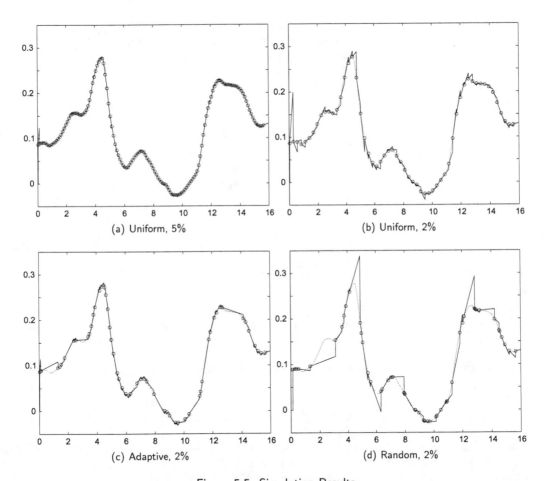

Figure 5.5: Simulation Results

Three ways of data reduction at the source have been applied in order to reduce the data rate efficiently. Figure 5.5 shows some representative results. Here, the gray line corresponds to the 180 Hz sensor data stream (all samples), while the circles mark those samples that have actually been transmitted by WSDP (updates). The black line represents the results at the receiver, i.e. the value delivered by the periodic WSDP-DATA.indications. In all cases, a simple first order TAYLOR approximation has been used, $\sigma^2 = 20$, $\tau^2 = 0.05$, and $T = 10$ ms. It is remarkable, how quickly start-up transients disappear. This fact gives an idea of the filter's adaptive performance.

In 5.5(a) data has simply been sub-sampled equidistantly, keeping only 144 of the original 2873 samples. This corresponds to 5% of the original data. For this specific application, obviously 95% of the sensor data can be easily dropped – there is hardly any deviation from the original data. In 5.5(b), this approach has been carried to extremes, keeping only 2% of the original data stream. This low data rate is not meant to be actually useful; instead, differences between the three rate reduction methods become evident. In 5.5(d) the data has not been sub-sampled uniformly, but updates have been selected randomly, again keeping only 2% of the data. In 5.5(c), the server-side KALMAN filter approach has been used to adaptively perform update selection. Again, only 2% (56 samples) of the available data are used as updates. But here, these samples have been selected adaptively, as outlined in section 5.5.4. However, the server-side filter runs at a much higher frequency, 1000 Hz compared to the 100 Hz of the client filter.

The adaptive scheme is best used when data transmission is reliable, since it identifies those updates that are most crucial for reconstruction. However, the threshold provides a means to exchange reliability against traffic reduction. Furthermore, the adaptive scheme could be extended in such a way that it transmits update packets with a minimum periodicity regardless of signal dynamics.

Regarding reliability, we can observe that even complete loss of update messages does hardly affect the quality of sensor value estimations – provided that a sufficient redundancy has been added. For instance, when we choose to reduce the data rate down to 10%, this would still allow us to loose an average 50% of the update messages, while estimation quality would still remain good. The effects of bursty transmission errors may be studied analyzing the results of random reduction presented in figure 5.5(d). If those updates, which were lost due to a corrupted burst, carry information crucial to successful reconstruction, prediction errors may become non-negligible. Refer to the situation around the turning point at $t = 4\,\mathrm{sec}$. In general, if sensor signal dynamics do not change dramatically during erroneous bursts, this will do only marginal harm to signal reconstruction. Furthermore, when receiving a valid update, the filter adapts quite fast, usually within a single cycle.

In figure 5.6 results are presented for a case of three sources observing the same phenomenon and a single intermediate station producing a set of updates to be forwarded towards the sink. In all sub-figures, the dotted line corresponds to the reference signal (all 2873 samples). Circles mark updates, while the solid line represents the KALMAN filter output. Sub-figures 5.6(a), 5.6(b), and 5.6(c) show the updates received by the intermediate station, originating in the individual sources A, B, and C, respectively. Notice that from 2873 samples, only an average 1% has been transmitted by the individual sources based on a purely random selection. The results of data aggregation are demonstrated in 5.6(d). Here, the filter output could be immediately made available to client applications if the intermediate node was also a sink. In 5.6(e) the adaptive scheme referred to earlier has been applied to the filter output (which is a 100 Hz reconstruction of the original sequence). This, in turn, yields the observations effectively forwarded by the intermediate station. In this case, approximately 1.6% of the original data are sufficient to reconstruct the signal.

5.9   Conclusion

Having identified some major problems associated with the transmission of sensor data over wireless networks, a new application layer protocol mitigating these deficiencies has been presented. It features ready-to-use, reliable and efficient data distribution across meshed ad hoc networks,

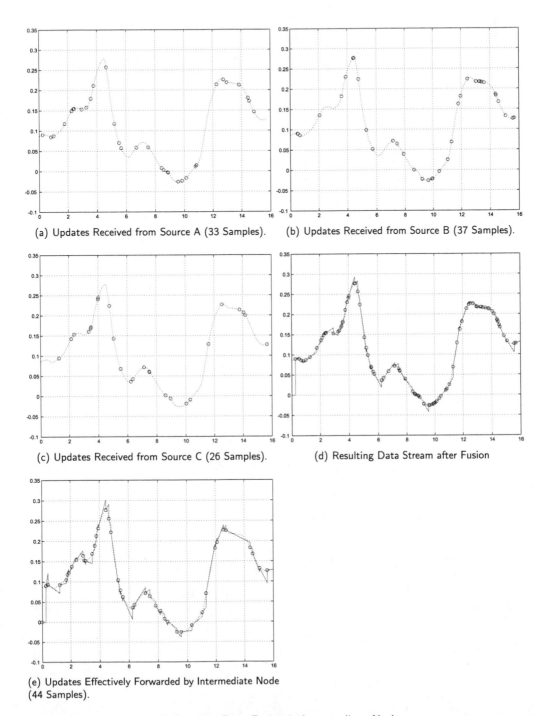

(a) Updates Received from Source A (33 Samples).

(b) Updates Received from Source B (37 Samples).

(c) Updates Received from Source C (26 Samples).

(d) Resulting Data Stream after Fusion

(e) Updates Effectively Forwarded by Intermediate Node
(44 Samples).

Figure 5.6: Data Fusion in Intermediate Node

which are exposed to hostile wireless conditions. Since WSDP works on top of almost any transport layer protocol, a huge number of potential applications may be thought of. The proposed framework may be combined with additional measures in lower layers, which aim at more reliable multicasts. In addition, the aforementioned benefits come at relatively low cost. Bearing in mind that computational costs are generally negligible compared to transmission costs in wireless sensor networks, adaptive filtering at transmitters and receivers takes full advantage of this notion. It has been argued that this kind of filter is a natural choice when it comes to implementing the advanced concept of data fusion in wireless sensor networks. Furthermore, the underlying mathematics, i.e. a KALMAN filter combined with a low order TAYLOR approximation, behaves very well with respect to numerical stability and robustness. The proposed solution is real-time capable, since the filter adds almost no latency.

Part II

Simulator Design

This page intentionally left blank.

CHAPTER 6

The Wireless Wide Area Network Simulator

In the previous part, a number of simulation results have been presented. These have been obtained with the Wireless Wide Area Network Simulator (WWANS), a new simulation environment, which has particularly been designed for system level verification and optimization of large scale ad hoc networks. Channel assignments obtained by TBSD-RRM, example routes determined by AODV, topology graphs, etc. all have been created using WWANS. This new simulator borrows some ideas from existing simulators like ns-2 and GloMoSim – such as the underlying discrete-event (DE) paradigm widely approved for network simulation – but is also substantially different in many respects. Starting with the first line of C++ code, back in 2001, WWANS has been designed to integrate with the hardware demonstrator platform to be introduced in part III. As a result, protocols once developed and tested in WWANS are instantly operational in the demonstrator – without any change. Part of this chapter has been published in [106–108].

6.1   Introduction

The question may arise: Yet another simulator? Why not use existing simulators, such as the frequently used ns-2, GloMoSim or commercial products for these purposes? There are a number of answers to this question, with one being of outstanding importance: The same code base shall be used for simulation, and – without modification – on real network devices, such as the SDR platform presented in part III.

6.1.1   Deficiencies of Existing Simulators

SDL and FSM based Simulators

Some simulators enforce the use of proprietary languages and structures to describe simulation models. For example, OPNET Modeler, a leading commercial product, requires models to be defined as finite state machines (FSMs). Other products often use SDL[1] specifications for automatic generation of C/C++ code, which is subsequently linked with a simulation kernel and existing models to conduct simulation runs. Usually, this is done to hide the pitfalls of parallel

---

[1]The specification and description language (SDL) has been designed by ITU-T for the specification of complex, event-driven, real-time, interactive systems

processing from designers. But experience tells that quality of generated code is generally poor, compared to optimized, hand-crafted code. In addition, it is almost unreadable for human beings.

GloMoSim

GloMoSim [109] is a simulation software, which builds on top of the Parallel Simulation Environment for Complex Systems (PARSEC). This C-based simulation language is suitable for sequential and parallel execution of DE simulation models. It can also be used as a parallel programming language. GloMoSim is a collection of PARSEC modules for the simulation of wireless networks. It provides a number of models covering mobility patterns, radio propagation, as well as PHY, MAC (including IEEE 802.11), network (including IP & AODV), transport (TCP & UDP), and application layers. However, PARSEC requires a cross-compiler translating PARSEC programs into C programs, which are finally linked with PARSEC's runtime kernel library. Furthermore, PARSEC does not support C++, which is a major drawback. Implementation of the IEEE 802.11 MAC layer is rather simplistic, albeit quality is superior to that of ns-2. As for ns-2, fragmentation and lifetime counters are not supported.

The Network Simulator ns-2

The Network Simulator, currently in its version ns-2 [50], is a discrete-event simulator targeted at networking research. It provides substantial support for simulation of TCP, routing, and multicast protocols over wired and wireless (local and satellite) networks. Originally, it had been designed for wired networks, though. This simulator is often used by researchers studying the IEEE 802.11 MAC layer, and a variety of other protocols for ad hoc and sensor networks. A large number of protocols, for example MPLS, streaming media, etc. have been provided by numerous third-party contributors.

In addition, ns-2 requires a wealth of other software and libraries for proper operation. It uses different languages, OTcl and C++, in combination. Simulation setup scripts are processed by the OTcl interpreter, which parameterizes the underlying simulation models written in C++. Quality of ns-2's IEEE 802.11 MAC layer model is rather poor. Apart from missing features (like fragmentation and lifetime counters), even late versions contained a number of bugs. For example, when a broadcast packet was passed to ns-2's IEEE 802.11 MAC implementation, it did not employ collision avoidance properly. Furthermore, ns-2 does not adhere to the IEEE 802.11 standard with respect to frame structures, service primitive interfaces, and so forth.

6.1.2   A New and Different Approach

Often MAC and PHY simulation code is merged together and the code is hardly usable outside the host simulator. This is different in WWANS, which has been written based on object oriented design paradigms. The simulator is modular in structure, facilitating reuse of once written and thoroughly tested code at different places within the simulation environment. Advanced language features unique to C++ and the standard template library (STL) have been extensively used to obtain elegant, though readable, high-performance code.

Conceptually isolated entities, for example PHY and MAC, are strictly separated. However, this does not prohibit cross-layer optimized architectures, when interfaces (service access points) are properly designed. The concept of abstraction is used to hide platform specifics from protocols and other subsystems that might be useful in simulator and demonstrator equally well. For example, timers are vastly used in a variety of protocols. In simulations, these timers are based on simulation time, perhaps in conjunction with a drifting clock model. On real network devices,

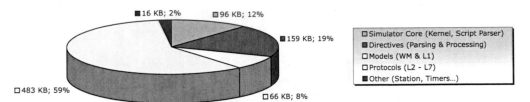

Figure 6.1: Source Code Structure of WWANS – Overview

timers are implemented by quartz-driven counters. To the protocol it must make no difference if its timer event handler is executed because the corresponding simulation time has been reached, or because a timer interrupt has occurred. Simulator/demonstrator integration, being a topic of its own, is beyond the scope of this chapter; it will be detailed in chapter 10.

The overall simulation environment is constituted by a number of executables, where wwans is the main simulation engine. It is accompanied by tools for visualization of graphs, creation of setup scripts describing random topologies, automatic generation of link activation requests, evaluation of assignments obtained by TBSD-RRM, etc. Some of these tools will be covered in the next chapter, where a number of typical simulation flows is going to be presented.

Just to give an impression of the simulator's complexity, a set of 211 C++ classes currently constitutes the main simulator engine, wwans. Each class contains a number of properties (member variables) and methods (member functions). Some classes are quite simple, for example those that define the binary structure of packets, but others contain complex implementations of TBSD-RRM, WSDP, IP and AODV protocols, for example. In addition to these classes, there are also a number of global functions, which have not been listed individually. What can be seen from this structure is the partitioning into easily reusable modules.

In figure 6.1 an analytical overview of the source code structure is presented. In total, the simulator currently consists of approximately 900 KB of hand-crafted, high-quality, compact C++ source code (including comments). No code generators have been used, no third party tools or libraries are required – except for the STL, which is provided by almost any compiler suite. It can be seen that the core (kernel, general script processing, and other management tasks) constitutes approximately 12% of overall source code volume. Another 19% are related to directives, including the associated script processing functions, and the corresponding actions in controlling simulation flow. Wireless medium, noise accumulating PMD model and IEEE 802.11 PHY interface sum up to 8%. The wealth of 59% represents protocol implementations for L2-L7 protocols, and the good news is that this code base is reusable in the hardware demonstrator to a great extent.

It is worth analyzing these 59% of common code base in more detail, as it is done in figure 6.2. A great partition of 25% is related to IEEE 802.11 MAC FSMs, frame structures, etc. The related code is partially or completely reusable in the hardware demonstrator, depending on how much of the MAC layer will be implemented on a microprocessor structure (DSP, Xilinx MicroBlaze or other softcore, embedded PowerPC in Xilinx Virtex-II Pro, etc.); and how much using dedicated circuitry developed in VHDL. All other protocols are completely reusable, one-to-one: Remember that directives and other simulation-specific structures have been separated from core protocols. This chart also allows a comparison of complexity. For example, NDP is quite a simple protocol (4%), while TBSD-RRM is rather complex (29%), as it contains UxDMA plus extensions, the transaction-based voting service, soft-decision table fusion, etc.

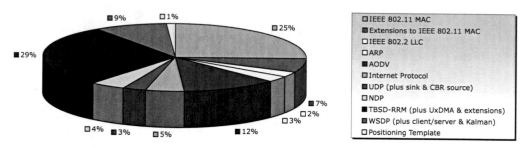

Figure 6.2: Source Code Structure of WWANS – Relation of Protocol Implementations

6.2   Architectural Concept

The main simulator engine's fundamental architecture is illustrated in figure 6.3. A number of stations are arranged within three-dimensional (3D) space. Currently, there is no support for modelling the terrain. That is, there is only free space[2], and stations.

A variable number of protocol instances can be attached to each station, where protocol stacks must not necessarily be equally structured. For example, different stations can be attached (possibly multiple) copies of different application layer protocols, etc. Connectivity between stations is calculated dynamically at simulation run time, using the wireless medium (WM) model and a flexible radio model with a threshold on the CINR.

Simulation is event-driven. At startup, when the setup script is processed by the configuration parser, a number of initiator events are registered with the simulator's event set. These will be processed by a powerful DE kernel. In future versions, a parallel processing kernel could be used to reduce simulation run times on multiprocessor systems[3].

6.2.1   Discrete Event Kernel

The DE paradigm is often used in network simulators because it offers several advantages compared to continuous-time or fixed time stepping approaches. The simulation is exact, that is, there are no approximations in the time domain, and it is limited to the processing of a bound number of events. At the engine's core, a sequential DE kernel is responsible for controlling overall simulation flow. Simply put, events are stored in a queue, sorted by due time (in simulation time units).

For example, assume that the configuration script contained two initiator events: one to turn-on a certain station at simulation time 10 s, another one to turn-off the same station after 30 s. Then the former event would be on top of the simulation event set queue, while the latter would be at the bottom. The sequential DE kernel always picks the topmost event and executes the associated action, in this case turning on/off a specific network element.

Scheduler

The scheduler, in conjunction with the associated event set queue, forms a DE kernel, which sequentially executes time-stamped events in appropriate order. Some events are initially given

---

[2]Except for a conceptual ground plane, when the two-ray-ground radio propagation model is used
[3]This will also improve performance on recent "hyper-threading" machines

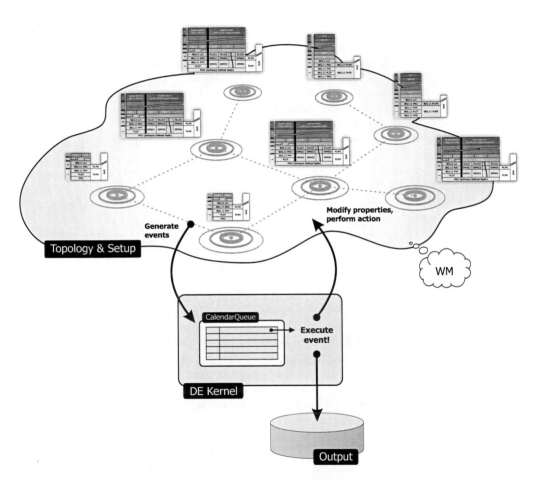

Figure 6.3: The Wireless Wide Area Simulator (Architecture)

in the configuration script, others are dynamically created as simulation advances. New events are queued according to their time stamp, with the earliest event always being on top of the pile. The scheduler picks the topmost event and executes all the related actions (i.e. it "consumes" the event), and, by doing so, advances the simulation time to the very instant identified by the event's time stamp. The simulation is complete, as soon as the scheduler's event queue is empty.

Event Queue

As already mentioned before, any action, manipulation of any property, and so forth are due to event handlers being executed. One can imagine that the number of events being processed during a simulation may be enormous. Every wireless transmission, for example, is modelled as a series of events: For each potential receiver, events are generated indicating the instants, (1) when the emitted electromagnetic wave arrives at the antenna, and, (2) when it has completely passed by. Obviously the size of the event set becomes tremendous, when a large number of stations are in the vicinity of the transmitter. The situation is even worse, whenever multiple transmissions are "on-the-air" simultaneously.

Consequently, efficient management of the event set is of major importance to simulators intended for dense, large-scale networks. Hence, a new C++ template class has been designed, realizing a so-called "calendar queue" [110], which is a fast implementation of a priority queue[4]. Its time complexity is $O(1)$ in, both, enqueue and dequeue operations. It should be emphasized that both operations are frequently performed. As an example, consider the case of timeouts: At many places within a station's protocol stack, from L2 to L7, timeouts must be observed. If a specific event does not occur in time, the associated timeout expires and the related protocol will learn that some operation has failed, for instance. But, when the event does occur in a timely fashion[5], the corresponding timeout event must be canceled again. Therefore, both operations are important.

The calendar queue achieves its performance by partitioning elements into a set of equally spaced bins. Then, bubble sort [95] is performed on the few elements in each individual bin. This is where the calendar queue got its name from, since its action is similar in operation to a calendar, where the set of days in a year is partitioned into months; one month for each page. Then, each bin resembles a monthly-page containing days in a particular month.

The number of bins is dynamically adjusted to be less than the number of elements (events in the particular case), and greater than half of the number of elements. As a result, there are one to two elements in each bin on average. The spacing between bins is also chosen and maintained so that the distribution of the elements in the bins is as close to a uniform distribution as possible. Generally, the bin-maintenance algorithm tries to achieve an optimal trade-off between time-complexity and storage-space. It is executed whenever the event set's size changes by more than a factor of, say, two. Also, the size is required to sustain this threshold for a while to avoid unfruitful overhead when the queue size is volatile.

6.3   A Note on Different Builds

It is worth clarifying that two versions of WWANS can be generated out of the same source code base. A fast **release** build, which does not provide any output except for statistics and explicitly requested output, such as graphs representing TBSD-RRM channel assignments, local

---

[4]In fact, in early versions of WWANS the STL `priority_queue` had been used, instead
[5]This is the more likely case in any functional system

NDP neighborhoods, etc. Explicit output is requested using special configuration script directives, which will be described later on.

In addition to the release build, a **debug** version is also available. This version runs much slower than the release build, but it performs extensive memory checking, employs a lot of ASSERTions to make sure that prerequisite assumptions are fulfilled, etc. In addition, the debug version provides implicit textual output in the form of: "Now, I'm doing this; next I plan to do that...", so as to provide insight into actions, state changes, etc. In the future, it is intended to store such information in binary format, and use a graphical viewer to visualize the wealth of information in a more accessible fashion. Then filters would be defined, which enable such logging for the models of interest, while suppressing undesired output otherwise.

6.4 Simulation Setup

As already mentioned before, simulations are conducted by running the wwans executable, and passing it a configuration script, which defines the actual simulation setup. An example for such a file is printed in listing 6.1. A formal description of all configuration directives is not in the scope of this work, but the script is pretty much self-explaining. A C-like syntax structure is used for comments, identifiers, string literals, etc.

```
// config.txt : A sample configuration script for WWANS
//

topography
{
    station
    {
        position = (-550.0, 550.0, 1.2);
        orientation = (0, 0, 0);
        clock = drifting(100e-6);

        protocols
        {
            attach(ieee_802_11_phy(2.4e9, 11e6, 2000000));
            attach(ieee_802_11_mac("00-00-00-ff-00-01"));
            attach(ieee_802_2_llc);
            attach(ipv4("192.168.1.1", "255.255.255.0", "0.0.0.0"));
            attach(aodv);
            attach(udp);
            attach(mlme_ndp(60.0, 0.5, 3));
            attach(mlme_rrm);
        }
    }

    station
    {
        position = (550.0, 550.0, 1.2);
        orientation = (0, 0, 0);
        clock = drifting(45e-6);

        protocols
        {
            attach(ieee_802_11_phy(2.4e9, 11e6, 2000000));
            attach(ieee_802_11_mac("00-00-00-ff-00-02"));
            attach(ieee_802_2_llc);
            attach(ipv4("192.168.1.2", "255.255.255.0", "0.0.0.0"));
            attach(udp);
            attach(aodv);
            attach(mlme_ndp(60.0, 0.5, 3));
            attach(mlme_rrm);
        }
    }
```

```
station
{
    position = (0, 0.0, 1.4);
    orientation = (0, 0, 0);
    clock = drifting(65e-6);

    protocols
    {
        attach(ieee_802_11_phy(2.4e9, 11e6, 2000000));
        attach(ieee_802_11_mac("00-00-00-ff-00-03"));
        attach(ieee_802_2_llc);
        attach(ipv4("192.168.1.3", "255.255.255.0", "0.0.0.0"));
        attach(udp);
        attach(aodv);
        attach(mlme_ndp(60.0, 0.5, 3));
        attach(mlme_rrm);
    }
}
}

radiopropagation
{
    carrierthreshold = 1e-16;
}

rrm
{
    constraints = (err0 | ett0 | etr0 | etr1);
    target = edge;
    ordering = mnf;
}

applications
{
    cbr_source("192.168.1.3", 8100, "192.168.1.1", 8100, 10.0, 20.0, 0.5, 8192);
    cbr_source("192.168.1.1", 8101, "192.168.1.2", 8100, 10.0, 20.0, 0.5, 4096);
    udp_sink("192.168.1.1", 8100);
    udp_sink("192.168.1.2", 8100);
}

events
{
    // Turn on stations simulataneously
    power("00-00-00-ff-00-01", 1, 0.0);
    power("00-00-00-ff-00-02", 1, 0.0);
    power("00-00-00-ff-00-03", 1, 0.0);

    // Turn off stations simulataneously
    power("00-00-00-ff-00-01", 0, 30.0);
    power("00-00-00-ff-00-02", 0, 30.0);
    power("00-00-00-ff-00-03", 0, 30.0);
}
```

Listing 6.1: Example of a Configuration Script for WWANS

6.5   Wireless Medium Model

The WM model incorporated into WWANS provides the ability to model wireless transmissions in a very flexible and generic way. In contrast to many other simulators which provide a single wireless channel only, this new model supports an unlimited number of concurrent transmissions using TDMA, FDMA, CDMA, SDMA and any combination thereof. Transmissions are placed "on-the-air" by a radio transceiver, which is modelled as a device being capable of transmitting and receiving on a multitude of wireless channels simultaneously. Transceivers may have a single antenna or multiple antennas with fixed or adaptive three-dimensional (3D) radiation patterns.

6.5.1 Transmissions and Channels

In simulations, a wireless transmission is characterized by a number of properties. This set has been chosen, such that a wide set of multiple access schemes is supported, while computational load remains acceptable for large-scale simulations. In addition to the transmitter's position and orientation, the following elements are used to distinguish transmissions:

- The time instants where the transmission starts, t_0, and ends, t_1,

- the transmitter power P_{TX},

- and the wireless channel.

Wireless channels are identified by their frequency band m, time slot n, and code sequence z, which are used to match transmitters and receivers. Furthermore, some more physical properties of the channel are relevant:

- Its center frequency f_0, used to determine the path loss,

- its bandwidth B, used to calculate the thermal background noise, and

- its capacity C, used to calculate the transmission delay (the time required for a transmission to complete).

6.5.2 Radio Propagation

There are two major aspects of radio propagation that are of importance to the simulation models: The path loss along the transmission path from transmitter to receiver and the propagation delay, e.g. the time it takes an electromagnetic wave to travel from transmitter to receiver. Given the EUCLIDean distance d between transmitter and receiver, and the speed of light c, the delay τ may be expressed as

$$\tau = \frac{d}{c}. \tag{6.1}$$

At present, two very simple propagation models, namely the free space [111] and two-ray-ground [30] models, are built into the simulator, where the latter performs better for far distances. The free space model is the most simple one and assumes a single undisturbed LOS link between transmitter and receiver. In this case, the path gain G_P, i.e. the relation between received and emitted power, can be calculated as

$$G_P = \left(\frac{\lambda}{4\pi \cdot d} \right)^2, \tag{6.2}$$

where $\lambda = c/f_0$ denotes the wave length[6]. In the two-ray-ground model, two paths exist for a single transmission. The first is the LOS path, as mentioned before, and the second is the path when the wave is reflected by the ground plane. The transmission gain is then given by the formula

$$G_P = \left(\frac{h_{TX} \cdot h_{RX}}{d^2} \right)^2, \tag{6.3}$$

with h_{TX} and h_{RX} being the distance between ground plane and transmitter/receiver antenna. More sophisticated multipath fading models [112] may be added in future versions of the simulator.

---

[6]This is a simplification, since the center frequency is substituted for each frequency in the entire band

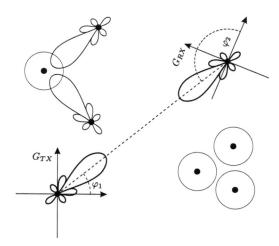

Figure 6.4: Potential Radiation Patterns (Example)

6.6   Models in Layer One

A noise-accumulating transceiver model (a PMD in IEEE 802.11 terms) is presented, together with a suitable PLCP which provides a subset of the IEEE 802.11 PHY service primitives, i.e. those primitives that are required by the MAC model for operation.

6.6.1   Transceiver Model

Antenna Radiation Patterns

In order to support different antennas, in particular smart or adaptive ones, radiation patterns may be specified. Antenna patterns are specified by their gain function $G(\vartheta, \varphi)$, and may be defined independently for the transmit and receive paths of a transceiver. Further on, they shall be represented by $G_{TX}(\vartheta_1, \varphi_1)$ and $G_{RX}(\vartheta_2, \varphi_2)$, respectively. An example is illustrated in figure 6.4. For any transmitter-receiver pair, the WM model calculates all four angles dynamically, thereby providing for node movement.

Support for FDMA

In order to calculate power reduction due to different frequency bands, a function $\mathcal{M}(m_1, m_2)$ is defined. It returns the factor by which the overall transmission loss is increased if a transmission at frequency band m_1 arrives at a receiver that is tuned to frequency band m_2. This is generally a function of the separation distance between the two bands, e.g. neighboring bands may cause significant adjacent channel interference, whereas frequency bands, which are separated by a wide gap, will not interfere at all. If the bands are identical no reduction occurs, i.e.

$$\mathcal{M}(m_1, m_2) = \begin{cases} 1 & \text{for } m_1 = m_2 \\ M(m_1, m_2) \leq 1 & \text{else} \end{cases}. \tag{6.4}$$

However, this function is implementation dependent, that is, it depends on the filters and components actually used, and the width of the guard band that separates neighboring bands. When the communication system employs a well designed radio resource management, the guard band's width may be reduced, leading to an increased spectral efficiency.

Support for CDMA

As we treat channels with different codes as separate physical channels, the effect of spread spectrum is taken into account at the transmission level. Similar to the function $\mathcal{M}$, a function $\mathcal{C}(z_1, z_2)$ is defined. If the receiver's code z_2 matches the transmitter's code z_1, the received signal power (after correlation with the spreading code) is the same as the power before application of the spreading code. Otherwise the received power is weakened by N, where N represents the spreading factor, or more formally:

$$\mathcal{C}(z_1, z_2) = \begin{cases} 1 & \text{for } z_1 = z_2 \\ \frac{1}{N} & \text{else} \end{cases} \tag{6.5}$$

Putting it all together

A transmission that is emitted at t_0 will arrive at the receiver at $t_0 + \tau$ and disappear at $t_1 + \tau$. The received power P_{RX} will equal

$$P_{RX} = P_{TX} \cdot G_{TX}(\vartheta_1, \varphi_1) \cdot G_P \cdot G_{RX}(\vartheta_2, \varphi_2)$$
$$\cdot \mathcal{M}(m_1, m_2) \cdot \mathcal{C}(z_1, z_2). \tag{6.6}$$

Notice that polarization has not been taken into account, since it would require LOS links to actually make use of this feature. In ad hoc networks, the majority of links are NLOS, thus the receiver will not be able to separate the originally differently polarized transmission signals, due to reflection and refraction.

However, this architecture allows for all the major multiple access schemes to be simulated correctly, e.g. TDMA, FDMA, CDMA, SDMA and any combination thereof (for instance multi-carrier TDMA as in GSM) may be used in the MAC layer. But it is also possible to use none of these schemes at all, leading to a single shared channel, as it is used by IEEE 802.11's CSMA/CA.

Interference

In principle, we can identify two kinds of interference, namely co-channel interference and adjacent channel interference. The first one is caused by co-located transmitters that transmit on the same frequency band, whereas adjacent channel interference is caused by non-ideal bandpass filters, imperfections of electronic components used in the receiver (intermodulation), etc. Usually, simulators for wireless ad hoc networks work with a single channel, and therefore neglect adjacent channel interference completely.

For CDMA systems, multiple access interference (MAI) is a major topic, since it is the limiting factor for dense node topologies and the driving force for transmitter power control. Here, every transmitter with a different code than the receiver contributes to the total interference (consider equation 6.5).

6.6.2 Thermal Noise

The thermal noise power P_N for a channel is calculated according to this formula:

$$P_N = kTB, \tag{6.7}$$

where k denotes the BOLTZMANN constant and T denotes the temperature, e.g. $T = 290K$. The background noise level is the limiting factor in the absence of interference.

Reception/Demodulation

Whenever an electromagnetic wave arrives at a transceiver, a decision must be made, whether the transmission can be successfully demodulated at all. This decision is made based on the carrier-to-interference-plus-noise ratio (CINR),

$$\text{CINR} = 10 \cdot \log \left(\frac{P_{RX,0}}{P_N + \sum_{i=1}^{K} P_{RX,i}} \right) \text{ dB},\tag{6.8}$$

where $P_{RX,0}$ denotes the received power of the desired transmission and $\sum P_{RX,i}$ accumulates the power of all concurrent transmissions (interferers) for a total of K transmissions.

A signal is said to be received successfully when the CINR is above a certain threshold. This threshold is implementation dependent, in that it depends on the receiver's noise figure, modulation scheme, etc. For all simulations carried out in the context of this thesis, a threshold CINR of at least 10 dB has been assumed.

In the future, the model could be further improved by introducing a stochastic bit error rate (BER) as a function of the CINR and the modulation scheme used, together with an algorithm which decides how much of these errors could be removed by an FEC stage. This way, some data packets will be corrupted and need to be retransmitted, for instance. More than that, the robustness of higher layer algorithms in the presence of corrupted or missing data frames may be analyzed.

Event Generation

Whenever the PHY needs to place a transmission "on-the-air", it does so by issuing a `WM-TXSTART.request` toward the WM model. The WM model then generates a number of new events and pushes them on the scheduler's event queue. The first kind of event, `WM-TXEND.indication`, is created once for each new transmission and is scheduled for the time the transmission ends (at the transmitter). This time can easily be determined, since the number of bytes (octets) to be transmitted and the channel's capacity are known entities.

Two other kinds of events are generated once for each receiver within a certain range. One event, the `WM-RXSTART.indication`, identifies the instant when the wave front reaches the receiver, while the other, the `WM-RXEND.indication`, denotes the time when the entire wave has passed the receiver. Since these values are different for different receiver locations, two events per receiver are required.

Link Closure

To limit the number of `WM-RXSTART.indication` and `WM-RXEND.indication` events, a link closure is applied. Both events are only generated when the carrier is above a certain threshold (at the receiver's site). This threshold can be adjusted in the configuration script, and allows a trade-off between accuracy and computational load, i.e. simulation time. Generally, if the carrier's power is only a fraction of the thermal background noise power, thermal noise will dominate – even if a number of interferers need to be taken into account. The following command in the configuration script's `radiopropagation` section may be used to specify this threshold:

```
carrierthreshold = 1e-16;
```

6.6.3 IEEE 802.11 PHY

In the preceding sections a WM model and some general techniques for creating a flexible radio model have been explained. Based on this, a noise accumulating model for the IEEE 802.11 PHY [10] tailored to WWANS shall be presented. Basically, thermal background noise plus interference (co- and adjacent channel) caused by concurrent transmissions are accumulated, and a threshold on the resulting CINR is used to decide whether reception is said to be successful, or not. This implementation provides the set of service primitives outlined in the protocol reference model for the IEEE 802.11 architecture. The PHY is responsible for the transmission of MPDUs over the WM. In WWANS an instance of the IEEE 802.11 PHY may be attached to any station. Using the following directive in the `protocols` section of a `station` declaration, the PHY is said to operate in the 2.4 GHz band at a data rate of 2 Mbps, and 11 MHz bandwidth (which corresponds to 22 MHz channel width):

```
attach(ieee_802_11_phy(2.4e9, 11e6, 2000000));
```

Whenever an MPDU has to be transmitted, a `PHY-TXSTART.request` is issued by the MAC. Upon reception of this request, the PHY will tune its transmitter to the specified channel and notify the MAC by means of a `PHY-TXSTART.confirm` that it is ready to receive the actual data. The MAC will then respond with a `PHY-DATA.request` carrying the information to transmit, which is acknowledged by the PHY via a `PHY-DATA.confirm`. A `WM-TXSTART.request` is issued, as soon as the data becomes available to the PHY. In the IEEE 802.11 reference model, for each octet of data one `PHY-DATA.request`, `PHY-DATA.confirm` cycle is required, i.e. the transmission is performed byte-by-byte. In contrast to this, the present implementation transfers the entire MPDU, e.g. 2346 bytes or less, in a single step. This is for efficiency reasons, and should not affect the simulation results. When the transmission is finished, the MAC issues a `PHY-TXEND.request` which is then answered by a `PHY-TXEND.confirm`.

Upon reception of a `WM-RXSTART.indication`, the PHY determines whether the total received power exceeds the carrier sense (CS) threshold. If so, the channel is marked as busy else it is said to be idle. As soon as a transition from idle to busy, or vice versa, is detected, the PHY informs the MAC layer of the change by the means of a `PHY-CCA.indication`. The clear channel assessment (CCA) state is cleared if – upon reception of a `WM-RXEND.indication` – the power level drops below the CS threshold. The MAC layer issues a `PHY-CCARESET.request` whenever the network allocation vector (NAV) of the virtual CS mechanism expires, in order to reset the CCA state machine, and a `PHY-CCARESET.confirm` is returned to the MAC.

Furthermore, if a carrier is available, which might be successfully demodulated (i.e. CINR $\geq$ 10 dB), a `PHY-RXSTART.indication` tells the MAC layer that a data packet is arriving. A `PHY-DATA.indication` transfers the received data to the MAC (again in one step), and via a `PHY-RXEND.indication` the MAC is informed that reception has completed. If the CINR drops below 10 dB during the time span, which is delimited by the `WM-RXSTART.indication` and `WM-RXEND.indication`, the received frame is said to be corrupted and a "carrier lost" condition is reported to the MAC. Otherwise, the frame is said to have been received successfully, and may be further processed by the MAC.

6.7 Protocols in Layer Two

Currently, L2 protocols embrace the IEEE 802.11 MAC, 802.2 LLC, as well as the innovative NDP and TBSD-RRM. In addition, a positioning protocol template is mentioned as a work-in-progress.

6.7.1 Medium Access Control

Each station, which provides a IEEE 802.11 PHY instance, can also be attached a copy of the IEEE 802.11 MAC. This can be done with the following directive, which is also used to set the station's unique hardware address:

```
attach(ieee_802_11_mac("00-00-00-ff-00-01"));
```

The new model presented in this section differs from other implementations in several ways. These differences may be subsumed into a single statement: The WWANS simulation model of the IEEE 802.11 DCF has been developed closely after the standard. Furthermore, the implementation has not been generated from an SDL specification, but hand-crafted C++ code has been developed instead. However, the SDL specification available from IEEE's 802 group has been used to clarify certain vague issues that were not covered comprehensively in the standards text.

Introduction

The IEEE 802.11 standard [10] together with its amendments − the most widely accepted standard for wireless local area networks − consists of several major building blocks. The most fundamental is the DCF, also known as carrier sense multiple access with collision avoidance (CSMA/CA), whose implementation into WWANS is detailed in this section. The DCF provides a basic contention service suitable for ad hoc peer-to-peer communications (P2P), and is also the basis for a contention-free service. The contention-free service, however, requires an access point to run the so called point coordination function (PCF) − essentially a master-slave polling scheme. Since we are primarily interested in the DCF as a means to realize a signalling service for future ad hoc networks, the PCF has not been implemented. Also, the MAC layer management entity (MLME) has been implemented only partially.

At the first glance, the question could arise why none of the existing simulation models for the IEEE 802.11 MAC have simply been adopted, for instance those that are provided in network simulators like ns-2 [50] or GloMoSim [109]. For a number of reasons, a new implementation was required, mainly because it is intended to use the IEEE 802.11 MAC protocol as a signalling path for future multi-channel wireless ad hoc networking stations. In parallel to the simulation software, a hardware demonstrator testbed (which is going to be presented in part III) has been developed, too. The goal is to use the same C++ code for both, simulations and real networking stations. This would not be feasible with existing implementations, since they tend to be simplistic, incomplete, and, with respect to some details, inaccurate.

Service Primitives

In ISO/OSI terminology, the MAC essentially provides two basic service primitives, namely:

- `MA-UNITDATA.request`, and

- `MA-UNITDATA.indication`.

These two primitives constitute the interface between MAC and LLC. This interface represents the upper boundary when regarding the protocol stack, while the interface to the 802.11 PHY forms the lower one. The new implementation presented here is tightly modelled after the standard, and therefore 802.11 MAC, 802.11 PHY and the underlying WM model are strictly separated. Other simulators, such as ns-2 and GloMoSim, often merge MAC, PHY and WM models into a single unit, making the implementations hardly reusable, and thus worthless for a hardware implementation.

6.7.2   Implementation Architecture

The MAC simulation model has mainly been realized using three finite state machines (FSMs):

- One for transmission (TX-FSM),

- one for reception (RX-FSM),

- and one for clear channel assessment (CCA-FSM).

Furthermore, the same binary frame structure for MAC protocol data units (MPDUs) is used as it is defined by the standard.

MA-UNITDATA.request requires a number of parameters, namely source and destination addresses, routing information, payload, i.e. MAC Service Data Unit (MSDU), priority and service class. In the present implementation, contention-free priority is not supported, since it would require the PCF. Also, only the **strictly ordered** service class is available, which prevents MSDUs from being reordered. Anyway, this does not really impose a restriction for the purposes of ad hoc network simulation. MA-UNITDATA.request will construct a valid 802.11 MAC header and pass it along with the payload to the TX-FSM.

Transmission Finite State Machine

Queueing of MPDUs   The TX-FSM, in turn, decides whether the MSDU should be fragmented, and if so, splits the MSDU into several fragments, where the fragmentation threshold is a configurable parameter. Each fragment forms an MPDU with its own header and trailer. The header carries address information, identifies fragments (fragment counter), distinguishes MSDUs (sequence counter), etc. This allows rejection of duplicate frames, enforcement of correct fragment ordering, etc. In addition, the header's duration field, which is used in the virtual carrier sense mechanism employed by CSMA/CA, is also calculated. The trailer would usually contain the frame check sequence, a 32-bit CRC value. Currently, this value is neither calculated at the transmitter nor evaluated at the receiver, in order to reduce computational load during simulation runs. A constant value is written in place of a meaningful CRC. A transmit lifetime counter is started, which, when expired, causes MSDU fragments that have not yet been successfully delivered to be dropped from the outstanding transmissions queue. This is useful, for instance, when a fragment burst has been interrupted for several times and multiple fragments needed retransmissions. In this case, it is up to higher layer protocols to invoke retransmission at a later time. Furthermore, if the MSDU length exceeds the RTS/CTS threshold, the TX-FSM creates a short request-to-send (RTS) frame, which has basically the effect of reserving the medium via virtual carrier sense for pending fragments. Also, the clear-to-send (CTS) frame which is transmitted in response to an RTS efficiently combats the hidden-terminal problem [32]. Finally, the FSM transitions to its next state, **carrier sense phase I**.

Physical carrier sense   Phase I of the carrier sense mechanism waits until the wireless medium becomes physically idle, i.e. the CCA does not detect any on-going transmissions. As soon as the medium becomes idle, TX-FSM enters the next state, **carrier sense phase II**. The task of this phase is to ensure that the medium remains idle for a certain amount of time. This time span, the inter-frame space (IFS), depends on several conditions; a thorough discussion of whom is beyond the scope of this thesis. In essence, the IFS allows for prioritization of different kinds of frames. Stations that currently obey to a short IFS are allowed to access the medium with a higher probability than those with a longer one. This way, ACK and CTS frames take precedence over outstanding transmissions of other competing stations, since they are associated with the shortest IFS available, SIFS. The IFS usually used in DCF mode is called DIFS, which is a little longer than SIFS. When the medium has not become busy again during the IFS, the station is allowed to continue its transmission process. To be accurate, two more conditions must also be met in order to proceed with the transmission: The virtual carrier sense must also indicate the idle condition, and the random back-off counter must have reached zero. Both mechanisms are outlined in the following paragraphs.

Random back-off   The random back-off procedure may be invoked from many places of the implementation. For instance, it is called, whenever the medium becomes busy during carrier sense phase II. The purpose of a random back-off is to prevent multiple competing stations to synchronize on the idle medium condition. If all stations obeyed to the same IFS, they would all start their retransmissions at the same time. With a high probability, this would cause the retransmissions to collide again, thereby impeding communications. Therefore, when a station invokes its back-off procedure, a pseudo random value is calculated, and added to the IFS. The random value's upper bound depends on current network contention: It grows exponentially with network load. The station with the shortest IFS wins the competition and commences transmission, hereby blocking the medium against rivals. Fairness is enforced as follows: A station that has detected the medium to be idle starts decrementing its back-off counter. When the medium becomes busy during back-off, decreasing the back-off value is interrupted. It is continued again, when the medium has become idle for the usual IFS. The next time the back-off procedure is invoked, a new random value is assigned only if the back-off counter was zero, otherwise back-off continues decreasing the previous value.

Virtual Carrier Sense   Effectiveness of physical carrier sense suffers from those near-far phenomena, which have been described in section 1.5.2, namely the hidden and exposed terminal problems. Virtual carrier sense, as it is employed in CSMA/CA, is able to mitigate at least the hidden terminal problem. Every MPDU being transmitted carries a duration field in its frame header. This field allows **a priori** medium reservation for future transmissions. Any station that receives an MPDU – regardless whether it is addressed or not – examines the duration field, which basically determines the amount of time the medium shall be reserved. The station updates its **network allocation vector** (NAV) when the duration field is greater than the current NAV. The medium is said to be busy, as long as NAV has not expired. A number of rules are provided on how duration needs to be calculated for the different frame types, i.e. data frames, RTS, CTS and ACK frames. The new simulation model presented in this section fully obeys to all the rules stated in the standards text.

Retransmissions   Immediate positive ACKs are used to determine whether MPDUs have been delivered successfully. In other words, TX-FSM switches into a waiting state after transmission of an MPDU has completed. It waits for the RX-FSM to report either a positive ACK frame or any other frame it has received. If TX-FSM does not receive a positive ACK notification from RX-FSM within ACK timeout, transmission is said to have failed and the MPDU is scheduled for retransmission. The same is true, if RX-FSM received any other packet than ACK. This is potentially a source for duplicate frames to be received by some station. When a station has successfully received a fragment and answered with a positive ACK, the ACK may be lost due to interference. In this case, the transmitter sends the MPDU again, and the receiving station must identify this fragment as a duplicate and reject it – which it does with the help of fragment counters. The MSDU's retry counter is incremented upon retransmission and the entire MSDU is dropped upon reaching a certain retry limit. Two retry limits exist for frames of different lengths. One for short frames, another for longer ones. Upon successful delivery of an MPDU, the retry counter for its associated MSDU is reset. Based on the retry counter, the **contention window** (CW) is calculated. This value influences the upper bound on the station's random back-off counter. The more retries are currently required, the longer the station has to back-off before it attempts retransmission. In addition, the station will no more obey to the DIFS, but to the longer EIFS (extended IFS) instead. This allows stations facing good network conditions to keep their throughput at a high level, while stations currently suffering from heavy network load step back and leave capacity for those stations who can actually use it.

Reception Finite State Machine

Reception of MPDUs   The MAC is informed of incoming MPDUs through a sequence of PHY indications. When reception of an MPDU has been successful, the `PHY-RXEND.indication` reports a "no error" condition. However, it may happen that packets collide and thus, a "carrier lost" condition may be reported instead. RX-FSM maintains a map of MSDUs it is currently receiving. It combines the transmitter's address and sequence counter to form a unique key which allows identification of MSDUs. Associated with the key is a list of fragment MPDUs that have been received as part of the MSDU. When the last fragment has been received, RX-FSM issues a `MA-UNITDATA.indication` transferring the complete MSDU (after defragmentation) to the IEEE 802.2 LLC. Using the map, it is possible t receive multiple fragmented MSDUs from different transmitters simultaneously. In case of an MMPDU, the associated MLME service receives the packet instead. The RX-FSM also utilizes its MPDU map to reject duplicates, as already indicated above.

Generation of response frames   Certain frames require the RX-FSM to answer with special response frames. The receiver will place ACK and CTS frames into the station's TX-FSM queue as appropriate. TX-FSM will guarantee that these response frames take precedence over all other pending frames. Furthermore, response frames are transmitted immediately after a SIFS has expired – without any carrier sense at all.

Maintaining the Network Allocation Vector   The RX-FSM is also responsible for NAV maintenance. It examines the duration field of all MPDUs it eventually receives and updates the NAV accordingly. After NAV has expired, the virtual carrier sense mechanism indicates an idle medium. Special rules exist for durations indicated by RTS frames. Here, if reception start has not been detected within a certain time, NAV is allowed to be reset earlier than requested in the RTS frame.

FSM Interworking

Obviously the state machines described above cannot act on their own, but rely on each other. During transmission, TX-FSM transmits a fragment and then waits for RX-FSM to report an incoming ACK frame, before TX-FSM continues with the next fragment. On the other hand, the receiver must utilize TX-FSM in order have its response frame (ACK or CTS) actually put on the air. Furthermore, TX-FSM relies on CCA-FSM in order to determine whether the channel is available or not.

Management Information Base

Statistical information is gathered during normal MAC operation. This information is part of IEEE 802.11 management information base (MIB), which is generally used for network management purposes.   Information collected in each station includes, for example, the number of successfully transmitted fragments (`dot11TransmittedFragmentCount`), the number of missing ACKs (`dot11ACKFailureCount`), the number of multicast frames received (`dot11ReceivedMulticastFrameCount`), ... – just to mention a few.

The complete set of variables is stated in the standards text. Those important for analysis have been implemented into WWANS. Besides these read-only information fields, there are also system parameters, which are accessible through the MIB. These include the thresholds for fragmentation and the RTS/CTS mechanisms, retry limits for long and short frames, as well as lifetime limits for MSDU transmission and reception.

6.7.3   Neighborhood Discovery Protocol

Of course, WWANS contains an implementation of NDP. Both NDP and TBSD-RRM are implemented exactly in the way presented in chapter 4, so their operation is not going to be repeated here. Instead, some simulation specific aspects are detailed in this part. NDP can be attached to any station, but it requires the IEEE 802.11 MAC protocol to be attached to that station, too:

```
attach(mlme_ndp(60.0, 0.5, 3));
```

This directive can be used in the `protocols` section of a `station` declaration within the configuration script file. In the present example, $T = 60$ s, $\sigma = 0.5$ s, $m = 3$. In addition to the plain protocol extending the MLME, the simulator also provides directives to take snapshots of local NDP tables. Snapshots are stored on the hard drive and are subsequently opened by the graphical visualization tool, `gview`, for evaluation. For instance, the following directive causes WWANS to take a snapshot of a certain station (identified by its hardware address) at the time specified, and to store a textual description of the graph on the hard drive:

```
__ndp_storegraph("00-00-01-00-00-02", "ndp-graph.txt", 15.5);
```

By the way, this technique has been used to create the graphs, which have been illustrated in figure 4.7. However, `gview` has been used to transform the textual graph description (which is going to be detailed in the next section dealing with TBSD-RRM) into a more accessible visual representation. While probe requests are usually broadcast automatically when a station is turned on, a further directive allows manually triggering individual NDP instances to broadcast probe request frames at a specified time for in-depth analysis:

```
ndp_probe("00-00-01-00-00-02", 5.0);
```

6.7.4   Radio Resource Management

Naturally, WWANS also contains a sophisticated implementation of the TBSD-RRM protocol (see chapter 4), which integrates with IEEE 802.11 MAC and NDP. Some settings, which are of global relevance, are specified in a dedicated `rrm` section, within the configuration script. In the following example, the constraint set is specified as $C = \left\{ \mathrm{E}_{rr}^0, \mathrm{E}_{tt}^0, \mathrm{E}_{tr}^0, \mathrm{E}_{tr}^1 \right\}$, edges are the targets for color assignment, and the MNF heuristic is selected for coloring local graphs obtained through NDP:

```
rrm
{
    constraints = (err0 | ett0 | etr0 | etr1);
    target = edge;
    ordering = mnf;
}
```

Individual instances of this protocol are easily attached to any station's protocol stack, provided that at least MAC and NDP protocols are also instantiated:

```
attach(mlme_rrm);
```

When simulation has started, and stations have been powered-on, RRM instances must be started-up. Afterwards, dedicated channels may be acquired and released after use. These actions are also controllable via directives:

```
// Startup RRM after NDP had enough time to collect neighbor information
rrm_startup("00-00-01-00-00-02", 15.0);

// Acquire a channel from one station to another at t = 20.0 sec
rrm_acquire("00-00-01-00-00-02", "00-00-01-00-00-1f", 20.0);

// Release the same channel at t = 40.0 sec
rrm_release("00-00-01-00-00-02", "00-00-01-00-00-1f", 40.0);
```

In addition, there is also a function for taking a snapshot of the local RRM table, similar to the directive mentioned in the context of NDP. It is possible to include only active links into this snapshot, or all links. This technique is essential for the evaluation of assignment results.

```
__rrm_storegraph("00-00-01-00-00-02", "rrm-graph-02.txt", 25.0);
```

Such a textual graph description is presented in listing 6.2. Here, a set of vertices is declared and a set of unidirectional edges connecting two vertices. In a separate section, vertex and edge colors are defined. For TBSD-RRM, these colors correspond to channel assignments, for example. But graph descriptions are widely used in WWANS at different places and with different meanings. Local NDP and TBSD-RRM tables, network connectivity (under the assumption of no interference), routes determined by AODV, etc. are also stored as graph files. Graph files may also contain a further section, which holds information useful for visualization, e.g. vertex coordinates, colors and highlighting indications. Refer to section 7.2.3 for more details.

6.7.5 Positioning Protocol Template

A template for a positioning protocol has also been added to the simulator. This template, which also serves as a test bench for the novel DMF introduced in chapter 3, gathers the results of delay measurements to immediate neighbors. Some work has been done to implement a GPS-free positioning protocol based on a fully distributed concept [113], but this is clearly a work-in-progress, and only mentioned for the sake of comprehensive coverage.

6.7.6 IEEE 802.2 Logical Link Control

On top of the 802.11 MAC, an instance of the 802.2 LLC provides the glue to L3 protocols, such as the Internet Protocol, ARP etc. The 802.2 standard defines several types of operation. The WWANS implementation of 802.2 currently only supports type 1, unacknowledged connection-less service. This type of service is particularly suited for transmission of IP packets.

```
// This file has been generated by wwans

graph
{
    // vertices
    vertex(v000000010000);
    vertex(v000000010007);
    vertex(v000000010009);

    // edges
    edge(v000000010000,  v000000010007);
    edge(v000000010000,  v000000010009);
    edge(v000000010007,  v000000010000);
    edge(v000000010007,  v000000010009);
    edge(v000000010009,  v000000010000);
    edge(v000000010009,  v000000010007);
}

colors
{
    vertex(v000000010000) = 18;
    vertex(v000000010007) = 16;
    vertex(v000000010009) = 22;
    edge(v000000010000,  v000000010007) = 62;
    edge(v000000010000,  v000000010009) = 0;
    edge(v000000010007,  v000000010000) = 56;
    edge(v000000010007,  v000000010009) = 1;
    edge(v000000010009,  v000000010000) = 8;
    edge(v000000010009,  v000000010007) = 12;
}
```

Listing 6.2: Example of a Textual Graph Description

6.8   Protocols in Layer Three

At present, L3 comprises the necessary protocols to create and maintain routes, and utilize these routes for the transmission of IP packets. Switching is done in the IP layer. That is, actually implementing the advanced concept of L2 cut-through switching, which has been introduced in chapter 2, remains to be done.

6.8.1   Internet Protocol

The new IP [6, 8] implementation designed for WWANS seamlessly integrates into any station's protocol stack. An instance of IEEE 802.2 LLC is required as the link layer protocol responsible for forwarding of IP packets. In the configuration script, the IP address is passed to the newly created protocol instance, together with the network mask and default gateway address:

```
attach(ipv4("192.168.1.1", "255.255.255.0", "0.0.0.0"));
```

The IP protocol accepts IP packets, which are typically provided by TCP or UDP, and forwards them to the next hop en route to the packet's final destination. A routing protocol is required to provide such routes to IP. The WWANS implementation of IP can be operated with any suitable routing protocol. Therefore, a modular interface has been designed, which allows plugging-in AODV, DSDV, TORA, static routing tables and so on. The routing protocol must provide a function, which returns the next hop's IP address given a packet's intended destination address. The next hop must be directly reachable through L2, i.e. it must be an immediate neighbor. In order to allow the routing protocol to keep track of actively used routes (for example to prolong route lifetimes), IP will notify the routing protocol, whenever it forwards a packet on a previously determined route.

When there is no route at hand, the packet remains queued until either its time-to-live (TTL) has expired, or a suitable route has become available in the meantime. The TTL is taken from the IP packet header, which must have been filled in properly by UDP, TCP, or any other entity directly passing IP packets to the Internet Protocol. The TTL is the time, in seconds, a packet is allowed to "survive" in the network. Each time, a packet is forwarded, the TTL must be decremented by one (at least). When a packet's TTL becomes zero, the packet is immediately discarded.

Assume that the TTL has not expired, and the next hop's IP address is known. Then the L2 hardware address is required for actual transmission of the IP packet over 802.2 LLC. To obtain the appropriate logical (IP) to hardware address mapping, ARP is employed. When the local ARP table contains a valid mapping, the packet is passed to L2, precisely following the "standard for transmission of IP packets over 802 networks" [71]. When no such mapping is instantly available, the packet remains in the queue, until the mapping eventually becomes available, or the TTL has expired.

6.8.2   Address Resolution Protocol

Mapping of logical to hardware addresses is done with the help of ARP. Again, WWANS provides a suitable implementation, which fully complies with standard requirements [70]. Notice that it is not necessary to attach the protcol using a configuration script directive. Instead, an instance of ARP is inherently attached to each IP instance. Nevertheless, a directive is available for manually triggering ARP requests. This directive can be added to the events section of the script:

```
arp_request("192.168.1.1", "192.168.1.93", 25.5);
```

ARP maintains a table, which stores the mappings between IP and hardware address spaces. If a mapping does not exist for any desired address, ARP broadcasts a request message, which basically contains four fields: the IP and hardware addresses of the sender (request initiator) and target.

Obviously, the sender knows its own logical and hardware addresses. Therefore, it can fill in these two fields. It also knows which IP address it is seeking, so it can fill in the target IP address field, too. When it broadcasts the packet, there is at most one station, which is assigned the target IP address. Hence, it creates a response packet. This time, it fills in the fields for sender IP and hardware addresses with its own address assignments, and uses the sender information from the request packet to fill in the target information for the response packet. Then, it broadcasts this response, and the originator of the request will have learned the desired address mapping.

In addition, all stations that are in transmission range will receive the broadcast, and extract useful information to be added to their own tables. To be more specific, every station receiving a request is able to retrieve the sender IP to hardware address mapping. Moreover, it can add both, sender and target address mappings of response frames. It should also be emphasized that the targeted station adds the sender's address mapping to its local ARP table, too.

It is worth noting that the ARP protocol also belongs to the class of protocols, which can benefit from the multicast scheme with improved reliability introduced in section 3.1. Instead of transmitting response frames as broadcasts, these could be immediately addressed to the originator of the request, which would add the benefit of reliable transmission from intended target back to the requester. In addition, letting all others listen to such directed response frames (even if not addressed), keeps efficiency of the ARP protocol at its original level.

6.8.3   Ad hoc On-demand Distance Vector Routing

As already mentioned before, the simulation environment also contains a novel implementation of AODV [46]. This protocol is intended for MANETs, as it offers quick adaptation to dynamic link conditions, low processing and memory footprints, and low network utilization to discover and maintain routes. Loop freedom is ensured at all times with the help of destination sequence numbers.

Overview

Compared to pro-active protocols, AODV is demand-driven. That is, it allows to obtain routes quickly for new destinations, whereas nodes are not required to maintain routes to destinations that are not in active communication. Whenever possible, local repair is used to respond to link breakages and network topology changes in a timely manner.

Three message types are defined by AODV, namely route requests (RREQs), route replies (RREPs) and route errors (RERRs). These messages are exchanged via UDP. AODV is a flooding-based protocol, that is, certain messages (such as RREQ) need to be disseminated widely – perhaps throughout the entire network. To limit the forwarding range for such messages, the IP packet header's TTL field is used.

Being a demand-driven protocol, AODV does not play any role as long as communication endpoints have valid routes. Only when a new route is needed, an RREQ is broadcast seeking the desired destination. As soon as the RREQ reaches either the desired destination or any intermediate note with a "fresh enough" route table entry, the identified route is made available to the requester by unicasting an RREP back to the originator of the RREQ. Nodes receiving an RREQ, cache the route back to its originator, so they are capable of forwarding a future RREP back along a path to that originator.

When a link break is detected, nodes use RERRs to notify other nodes en route about the lost link. An RERR message indicates those destinations that are no longer reachable due to the broken link. As a matter of fact, each station needs to know which of its immediate neighbors are likely to use it as a next hop toward those destinations that have become unreachable. Stations maintain a "precursor list" for this purpose.

In addition, AODV provides further sophisticated features very useful for its application to wireless ad hoc networks. For example, it supports operation over networks having unidirectional links using a "blacklist" set for next hops, to which forwarding of RREPs has failed. Moreover, the number of RREQ and RERR messages per time unit may be limited to, say, ten messages per second. Furthermore, network congestion is reduced by utilizing a binary exponential back-off (comparable to that of IEEE 802.11's DCF), in repeated RREQ attempts. Unnecessary network-wide flooding of RREQs is avoided by using an expanding ring search technique. At the beginning, a relatively low TTL value narrows the range of dissemination. If no RREP arrives within a certain timeout, the TTL is incremented to cover a greater distance, and the request is reattempted. A glance at AODV's route table entry structure gives an impression of its current and potential abilities. Each entry comprises...

- destination IP address,

- destination sequence number,

- state and routing flags (sequence number valid?, route valid?, . . .),

- distance (number of hops needed to reach destination),

- next hop's IP address,

- list of precursors, and

- lifetime (route expiration/deletion time).

More on the Implementation

Although there are number of lessons to learn from how AODV has been designed to cope with MANET challenges, a detailed discussion is well beyond the scope of this work. Here, AODV is just used as a vehicle for route determination. Once a route has been determined, it can be used for resource reservation. Dedicated resources may be allocated for active links using TBSD-RRM.

Three configuration script directives provided in the simulator are related to AODV. The first one, to be used in a station's protocol declaration section, attaches an instance of AODV to a protocol stack. In this case, AODV takes the role of the routing protocol for the associated IP instance. Notice that an UDP instance is also required, since AODV exchanges its messages via UDP:

```
attach(aodv);
```

Routes are usually requested by the IP protocol implementation for pending packets, as necessary. For testing and in-depth analysis, another directive (which may be used to test any routing protocol in future versions of the simulator) allows requesting a route manually:

```
route_request("192.168.1.10", "132.195.129.2", 75.0);
```

The last in the set of AODV related directives can be used to visualize determined routes with the help of gview. Once again, a textual graph description is stored on the hard disk, where source, destination and intermediate stations, together with related communication links appear highlighted:

```
__route_storegraph("192.168.1.10", "132.195.129.2", "route.txt", 76.0);
```

Finally, it should be stressed that the novel WWANS implementation of AODV is not only a simulation model, but is instantly reusable in the hardware demonstrator. Of course, the binary message structure for RREQs, RREPs and RERRs – as defined by the related RFC document cited at the beginning of this section – is closely followed.

6.9 Protocols in Layer Four

At present time, UDP is the only transport layer protocol provided in L4. In future work, the simulator may be extended by a TCP implementation, as well.

6.9.1 User Datagram Protocol

UDP [68] provides a thin wrapper around IP. It adds source and destination ports, which allow to differentiate between a multitude of applications. For example, AODV can exchange its messages through port 654, while another protocol, e.g. WSDP uses a different port for its messages. In order to reduce computational load, the current WWANS implementation of UDP neither calculates nor verifies CRC checksums. It is safe to do so, because the 802.11 PHY model does not forward corrupt frames. When interference is overwhelming, the PHY model reports a "carrier-lost" condition to the MAC, and the packet will never reach 802.2 LLC, IP, or even UDP. This might be different in future versions of the simulator (or the demonstrator), though. The same syntax as for the other protocols is used to attach an instance of UDP to any station's protocol stack. Anyway, an instance of IP is required.

```
attach(udp);
```

6.10   Protocols in Layer Seven

A number of application layer protocols are available for traffic generation, and transmission in wireless sensor networks. A CBR source generates UDP packets, which a generic sink accepts. Furthermore, the WSDP protocol introduced in chapter 5 has been implemented together with WSDP server and client applications. With their help, performance of the innovative KALMAN filter approach in sensor networks has been analyzed.

6.10.1   Constant Bit-rate Application

A constant bit-rate application is able to mimics the requirements of voice communications, for example. Generally, transport of CBR traffic is quite demanding. A network with reasonable QoS is expected to deliver this kind of traffic, such that inter-arrival times are regular, delay is bounded and a certain average throughput is sustained.

In configuration scripts, instances of the CBR source are declared in the `applications` section. Here the first IP address identifies the source, followed by the source UDP port, the destination IP address, and the destination UDP port. Additionally, the instants when the traffic generator is started and halted need to be specified, together with the inter-packet gap and the data size in bytes:

```
cbr_source("192.168.1.3", 8100, "192.168.1.1", 8100, 10.0, 20.0, 0.5, 8192);
```

Notice that this size corresponds to the size of user data to be transferred. UDP, IP, MAC etc. will add overhead according to their respective frame structures. It is possible to attach multiple source applications to the same station (using different source ports), which will run these instances in parallel.

6.10.2   Generic Sink for UDP Traffic

A generic sink application has been created, which accepts UDP datagrams on the specified UDP port. This application can be used to assess the network's ability to deliver packets in a timely fashion, detect violations of contracted QoS, etc. The sink can be used for UDP traffic generated by the CBR source, the WSDP server, or any other UDP-based protocol to be added in the future. The following directive installs a UDP sink application:

```
udp_sink("192.168.1.1", 8100);
```

6.10.3   Wireless Sensor Data Protocol

The novel simulator presented in this chapter, has already been successfully used during the design, verification and optimization cycle involved in the development of WSDP. The details of this protocol, its theoretical foundation and practical application have been provided in chapter 5. Therefore, this section does only provide a short description of the single configuration script directive immediately related to WSDP as such[7]. This directive is used to attach an instance of WSDP to a station's protocol stack, thereby telling WSDP which UDP port to use for data exchange with peer instances:

```
attach(wsdp(8500));
```

6.10.4   Wireless Sensor Data Server Application

The following directive may be specified in the configuration script's `applications` section to construct a WSDP server application:

```
wsdp_server("192.168.1.3", "255.255.255.255", 8086, 8.0, "sensor-updates.txt");
```

Besides source and destination IP addresses (here, a limited broadcast), a SID is used to distinguish different sensors. Further arguments to the directive specify the time, when the WSDP server is to be started, as well as a text file providing sensor readings. This file contains an arbitrary number of (time, value) pairs which emulate values that a sensor would provide in real-world applications. Notice that a separate tool, `wsdpprep`, is available for preparing an appropriate set of sensor updates for simulation given a reference value stream. This tool is described separately in section 7.4.

6.10.5   Wireless Sensor Data Client Application

Now that we have a WSDP server, a client application is required, which employs WSDP to create a continuous sensor value stream from the few updates transmitted by the server. Such an application is created just as easily as the server application. The following directive may be used to create the application and configure the underlying WSDP instance to deliver a stream of estimated of sensor readings with the desired sampling rate:

```
wsdp_client("192.168.1.1", 8.0, 24.0, 0.01, 8086, 2, "sensor-results.txt");
```

In this case, the IP address of the station intended to host the client is followed by the start and stop times for the WSDP subscription, the time period of the clock used to drive the KALMAN filter (here 10 ms, according to a 100 Hz synchronous output rate), the SID of the sensor in interest, the desired TAYLOR approximation order and the name of a text file that shall receive the reconstructed sensor stream. This file is of the same format as the one used with the server, i.e. a series of (time, value) pairs is recorded.

6.11   Conclusion

A novel simulation environment dedicated to future wireless multihop networks has been presented. While it would generally be possible to simulate all kinds of wireless (and wired) net-

---

[7]Additional directives exist to set up WSDP client and server applications

works, the current set of models and protocols focuses on MANETs. Although WWANS is a work-in-progress, it has already been successfully used by colleagues [114] and students [96, 113] for research. It is an important stand of a two-fold approach towards large-scale QoS-aware wireless multihop ad hoc networks.

## CHAPTER 7

Some Typical Simulation Flows

In the previous chapter, the fundamental simulator structure of WWANS has been outlined. This chapter considers some typical simulation flows, so as to figure out what kind of information can be drawn from simulations at all. By the end of this chapter, it shall have become evident that WWANS is able to provide deep insight into all parts of the envisioned communications system. As a result, it is suitable for fine-grained simulations and analysis of cross-layer optimization approaches. Furthermore, some new tools and their way of interaction are introduced. These tools are useful for script file generation, automated evaluation of channel assignment results, detection of constraint set violations, and so on.

7.1   Introduction

The main simulation engine wwans, whose architecture has been sketched in chapter 6, is accompanied by a set of new tools. These mainly operate on the two kinds of files mentioned earlier, i.e. configuration script files and graph files. In the following sections, it shall be explained how simulations are usually conducted with WWANS. The same principles have been used to obtain the results for TBSD-RRM, and WSDP, for example. It also shows that WWANS is not only modular with respect to its internal object oriented class structure, but also with respect to its external interface to the "outside world": It is quite simple to add further tools in the future, which build upon the same file structures that are used for the set of tools presented here. This is due to the fact that C++ classes are available, which parse script and graph files, and export graph files based on their internal, easy to manipulate representation in memory.

7.2   First Example: A Single Three-hop CBR Flow

In the first example, a complete flow is presented. We will start with generating a random topology using the topgen tool. Next we will run wwans to obtain the connectivity graph based on the specified radio settings and propagation model. With this information, we will setup a data flow using a CBR source for traffic generation, and a UDP sink for reception. Then, we will watch the simulator run the various models and protocol functions ranging from said applications, over UDP, IP & AODV & ARP, to IEEE 802.2 LLC, 802.11 MAC, 802.11 PHY and finally WM.

7.2.1   Automatically Creating a Script File

While it is possible to use any text editor to create a configuration script file for WWANS, it may be quite tedious to create such a file from scratch. Hence, a tool called `topgen` is provided, which is intended to perform this task for us. The `topgen` console program is started like this:

```
topgen -n5 -x1000 -y1000 -z2 config.txt
```

This will instruct `topgen` to create a topology consisting of five stations randomly distributed in a ± 1000 m $\times$ ± 1000 m $\times$ 2 m cube. Notice that for each station's position $z > 0$ holds true, to avoid problems that could otherwise arise in conjunction with the two-ray-ground model, as the ground plane is situated at $z = 0$. In this case, `topgen` will create a configuration script similar the one presented in listing 7.1.

```
// config.txt
//
// This WWANS script has been generated by topgen

topography
{
    station
    {
        protocols
        {
            attach(ieee_802_11_phy(2.4e9, 11e6, 2000000));
            attach(ieee_802_11_mac("00-00-00-01-00-00"));
            attach(ieee_802_2_llc);
            attach(ipv4("192.0.0.0", "255.0.0.0", "0.0.0.0"));
            attach(mlme_ndp(60.0, 0.5, 3));
            attach(mlme_rrm);
        }

        position = (835.383, -228.37, 1.04129);
        orientation = (0, 0, 0);
        clock = drifting(-5.25834e-006);
    }

    station
    {
        protocols
        {
            attach(ieee_802_11_phy(2.4e9, 11e6, 2000000));
            attach(ieee_802_11_mac("00-00-00-01-00-01"));
            attach(ieee_802_2_llc);
            attach(ipv4("192.0.0.1", "255.0.0.0", "0.0.0.0"));
            attach(mlme_ndp(60.0, 0.5, 3));
            attach(mlme_rrm);
        }

        position = (-490.524, 249.489, 3.99927);
        orientation = (0, 0, 0);
        clock = drifting(-1.48625e-006);
    }

    station
    {
        protocols
        {
            attach(ieee_802_11_phy(2.4e9, 11e6, 2000000));
            attach(ieee_802_11_mac("00-00-00-01-00-02"));
            attach(ieee_802_2_llc);
            attach(ipv4("192.0.0.2", "255.0.0.0", "0.0.0.0"));
            attach(mlme_ndp(60.0, 0.5, 3));
            attach(mlme_rrm);
        }

        position = (755.974, 252.663, 2.76485);
        orientation = (0, 0, 0);
        clock = drifting(8.71639e-005);
    }

    station
    {
        protocols
        {
            attach(ieee_802_11_phy(2.4e9, 11e6, 2000000));
            attach(ieee_802_11_mac("00-00-00-01-00-03"));
            attach(ieee_802_2_llc);
            attach(ipv4("192.0.0.3", "255.0.0.0", "0.0.0.0"));
            attach(mlme_ndp(60.0, 0.5, 3));
            attach(mlme_rrm);
        }

        position = (173.742, -296.426, 0.456069);
        orientation = (0, 0, 0);
        clock = drifting(-8.76034e-005);
    }
```

```
station
{
    protocols
    {
        attach(ieee_802_11_phy(2.4e9, 11e6, 2000000));
        attach(ieee_802_11_mac("00-00-00-01-00-04"));
        attach(ieee_802_2_llc);
        attach(ipv4("192.0.0.4", "255.0.0.0", "0.0.0.0"));
        attach(mlme_ndp(60.0, 0.5, 3));
        attach(mlme_rrm);
    }

    position = (-199.255, 830.134, 0.251717);
    orientation = (0, 0, 0);
    clock = drifting(7.17887e-005);
}
}

events
{
    // Power-on all stations
    power("00-00-00-01-00-00", 1, 0);
    power("00-00-00-01-00-01", 1, 0.5);
    power("00-00-00-01-00-02", 1, 1);
    power("00-00-00-01-00-03", 1, 1.5);
    power("00-00-00-01-00-04", 1, 2);

    // Put your events here

    //{{ddcagen(12.5, 0.5)}}

    // Store local RRM graphs
    __rrm_storegraph("00-00-00-01-00-00", "rrm-graph-v000000010000.txt", 7.5);
    __rrm_storegraph("00-00-00-01-00-01", "rrm-graph-v000000010001.txt", 7.5);
    __rrm_storegraph("00-00-00-01-00-02", "rrm-graph-v000000010002.txt", 7.5);
    __rrm_storegraph("00-00-00-01-00-03", "rrm-graph-v000000010003.txt", 7.5);
    __rrm_storegraph("00-00-00-01-00-04", "rrm-graph-v000000010004.txt", 7.5);

    // Power-off all stations
    power("00-00-00-01-00-00", 0, 12.5);
    power("00-00-00-01-00-01", 0, 12.5);
    power("00-00-00-01-00-02", 0, 12.5);
    power("00-00-00-01-00-03", 0, 12.5);
    power("00-00-00-01-00-04", 0, 12.5);
}
```

Listing 7.1: Configuration Script, as Generated by `topgen`

7.2.2 Running WWANS to Determine Connectivity

Given the configuration script, the question is: "what does the network topology graph look like?". Obviously this depends on a lot of factors, such as transmission power, the radio's noise figure, antenna gains, etc. The main simulator engine is able to process the configuration script and determine connectivity taking all these considerations into account. It will create a network graph, which describes stationary network connectivity in absence of interference:

```
wwans config.txt -gconnectivity.txt -x
```

Above command instructs the simulator to read the script file "config.txt", setup the stations in 3D-space, and determine connectivity based on predefined power settings (which could also be defined in the setup script). This path is future-proof: if, for example, in a future version of WWANS a terrain model is included, such features instantly affect connectivity. The trailing "-x" switch tells WWANS to skip the actual simulation, i.e. not to process any events. The resulting graph file, "connectivity.txt", output by `wwans` is shown in listing 7.2. Here, connectivity information is encapsulated in the `graph` section. The other two optional sections, `visualization` and `colors`, provide further information on top of connectivity alone. In this case, colors indicate a station's total degree. In addition, vertex positions are provided to simplify visualization.

```
// This file has been generated by wwans

graph
{
    // vertices
    vertex(v000000010000);
    vertex(v000000010001);
    vertex(v000000010002);
    vertex(v000000010003);
    vertex(v000000010004);

    // edges
    edge(v000000010000, v000000010002);
    edge(v000000010000, v000000010003);
    edge(v000000010001, v000000010003);
    edge(v000000010001, v000000010004);
    edge(v000000010002, v000000010000);
    edge(v000000010002, v000000010003);
    edge(v000000010003, v000000010000);
    edge(v000000010003, v000000010001);
    edge(v000000010003, v000000010002);
    edge(v000000010004, v000000010001);
}

visualization
{
    position(v000000010000) = (835.383, -228.37, 1.04129);
    position(v000000010001) = (-490.524, 249.489, 3.99927);
    position(v000000010002) = (755.974, 252.663, 2.76485);
    position(v000000010003) = (173.742, -296.426, 0.456069);
    position(v000000010004) = (-199.255, 830.134, 0.251717);
    comment(v000000010000) = "192.0.0.0";
    comment(v000000010001) = "192.0.0.1";
    comment(v000000010002) = "192.0.0.2";
    comment(v000000010003) = "192.0.0.3";
    comment(v000000010004) = "192.0.0.4";
}

colors
{
    vertex(v000000010000) = 4;
    vertex(v000000010001) = 4;
    vertex(v000000010002) = 4;
    vertex(v000000010003) = 6;
    vertex(v000000010004) = 2;
}
```

Listing 7.2: Connectivity Graph, as Determined by wwans

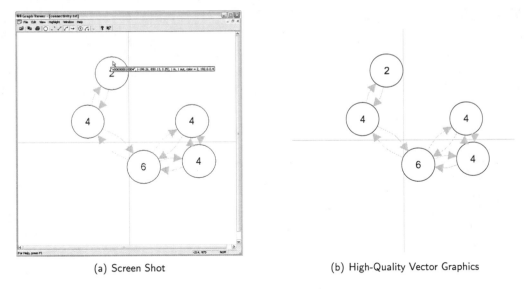

(a) Screen Shot   (b) High-Quality Vector Graphics

Figure 7.1: The Graph Viewer

7.2.3 Using the Graph Viewer for Visualization

The textual graph description is useful as input to a variety of algorithms, such as UxDMA (as we will see in the next section). Anyway, such an abstract description is inaccessible to the human mind, especially when the number of stations exceeds the few nodes of this simple example. Therefore, a further tool has been developed, which can be used to turn textual graph descriptions into their visual counterparts. These present a very natural and human-friendly form of graph representation.

Among the set of current WWANS tools, gview, the Graph Viewer, is the only one to employ a graphical user interface. While all other tools are instantly portable to a variety of operating systems[1], gview is intended for the Windows operating system alone. This is due to the fact that the Microsoft Foundation Classes library has been used, which offers a well-designed, professional application framework for C++ programs. The graphical viewer would allow us to fine-tune the configuration if we were not pleased with the outcome of topgen. For example, it would be feasible to create bottlenecks, bridge potential gaps, or even discard the topology altogether, and create another one.

In figure 7.1(a) a screen shot of gview displaying "connectivity.txt" is shown. From this figure, the purpose of the visualization section becomes evident. First of all, gview knows where to put each station using its x and y coordinates. Furthermore, tool tips appear dynamically as the mouse hovers over a vertex, presenting additional information (such as the IP address) which would be too confusing if shown simultaneously for all stations. The program is able to create scalable, high-quality vector graphics; a sample for the same graph is presented in figure 7.1(b). In fact, all graphs presented throughout this work have been created by gview.

---

[1]Versions for Linux and IRIX have already been built

7.2.4   Simulating a CBR Data Flow

Now that we are content with the topology, we want to make use of WWANS, in order to simulate a CBR data flow. The top-left station, which has the hardware address `00-00-00-01-00-04` and the IP address `192.0.0.4` sends UDP packets to the bottom-right station with hardware address `00-00-00-01-00-00`, and logical address `192.0.0.0`.

At the beginning, we need to modify the configuration script a little, so as to reflect the actual setup we want to simulate. First, NDP and TBSD-RRM, which have been added automatically by `topgen` are not required for this simulation and can be removed from the protocol stacks. Instead, UDP and AODV need to be added, yielding the following overall stack:

```
protocols
{
    attach(ieee_802_11_phy(2.4e9, 11e6, 2000000));
    attach(ieee_802_11_mac("00-00-00-01-00-02"));
    attach(ieee_802_2_llc);
    attach(ipv4("192.0.0.2", "255.0.0.0", "0.0.0.0"));
    attach(aodv);
    attach(udp);
}
```

Furthermore, we want to define two applications: a CBR traffic generator at the source, which starts at simulation time $t = 10$ s to transmit 1 KB of raw user data every 0.5 s until $t = 20$ s; and a mating UDP sink at the client station. Furthermore, we want to power-on all stations simultaneously at simulation time $t = 0$ s, and power them off after 25 seconds:

```
applications
{
    cbr_source("192.0.0.4", 8100, "192.0.0.0", 8100, 10.0, 20.0, 0.5, 1024);
    udp_sink("192.0.0.0", 8100);
}

events
{
    // Power-on all stations
    power("00-00-00-01-00-00", 1, 0);
    power("00-00-00-01-00-01", 1, 0);
    power("00-00-00-01-00-02", 1, 0);
    power("00-00-00-01-00-03", 1, 0);
    power("00-00-00-01-00-04", 1, 0);

    // Power-off all stations
    power("00-00-00-01-00-00", 0, 25.0);
    power("00-00-00-01-00-01", 0, 25.0);
    power("00-00-00-01-00-02", 0, 25.0);
    power("00-00-00-01-00-03", 0, 25.0);
    power("00-00-00-01-00-04", 0, 25.0);
}
```

On a 2.8 GHz Pentium IV machine, the release version of WWANS runs this simulation astonishingly fast: only 15 ms are required. However, as we have not requested any output data, there is also no obvious simulation result at hand. But we can ask the simulator to provide some of the statistics collected in the 802.11 MAC MIB by specifying a special "-s" switch on the command line. This will generate the output printed in listing 7.3 and reveals the desired statistics.

The debug build provides a lot more information. In fact, the resulting text file's size exceeds 1 MB for this scenario. On the same machine, 375 ms are required to create this output, which is again quite fast. In principle, every desired information can be included into this output. In addition, directives are easily added, which can be used in release builds as well.

Interesting action occurs at $t = 10$ s, when the CBR source starts transmitting packets. In

```
wwans -s config.txt

The Wireless Wide Area Network Simulator Version 1.0
Copyright(C) 2001 - 2004 University of Wuppertal, Germany. All rights reserved.
Reading configuration script "config.txt"...

Carrier threshold = 1e-016
Simulating topography with 5 stations...
Simulation completed in 0.015000 seconds.
```

station	frags (tx)	frags (rx)	failed	ACK errs	CTS errs	RTS/CTS ok	retries	seq. errs	mframes (tx)	mframes (rx)	dups
00-00-00-01-00-00	22	21	0	0	0	0	0	0	1	3	0
00-00-00-01-00-01	44	22	0	0	0	0	0	0	3	6	0
00-00-00-01-00-02	0	0	0	0	0	0	0	0	0	4	0
00-00-00-01-00-03	44	22	0	0	0	0	0	0	3	4	0
00-00-00-01-00-04	22	1	0	0	0	0	0	0	3	3	0

overall	frags (tx)	frags (rx)	failed	ACK errs	CTS errs	RTS/CTS ok	retries	seq. errs	mframes (tx)	mframes (rx)	dups
5	132	66	0	0	0	0	0	0	10	20	0

Listing 7.3: Output of wwans Release Build with "-s" Option

the following, it is assumed that the reader is familiar with the protocols being discussed, i.e. the various fields in packet headers, etc. are not going to be covered in detail. The debug version of wwans yields the following output for this time, with respect to the source `192.0.0.4`:

```
<000000010004> /CBR source/ <CBR>: issuing UDP-DATA.request (1024 bytes)
<000000010004> /UDP/ <UDP>: creating datagram: source = 8100, destination = 8100 (1032 bytes)
<000000010004> /UDP/ [UDP]: forwarding 1032 bytes to IP
<000000010004> /IPv4/ [IP]: IP-PACKET.request
<000000010004> /IPv4/ <IP>: +++++++++++++++ IP packet +++++++++++++++
<000000010004> /IPv4/ <IP>: ++ version = 4, IHL = 5, type of service = 0, total length = 1052
<000000010004> /IPv4/ <IP>: ++ identification = 0000, flags = 0, fragment offset = 0
<000000010004> /IPv4/ <IP>: ++ time to live = 255, protocol = 17, checksum = 0000
<000000010004> /IPv4/ <IP>: ++ source = "192.0.0.4", destination = "192.0.0.0"
<000000010004> /IPv4/ <IP>: +++++++++++++++++++++++++++++++++++++++++
```

At this point IP asks AODV to return the next hop toward the destination. This is when AODV recognizes that it has no suitable route at hand. Hence, it decides to initiate a route discovery and broadcasts an RREQ via UDP port 654:

```
<000000010004> /AODV/ [AODV]: route request
<000000010004> /AODV/ <AODV>: -------------- route request --------------
<000000010004> /AODV/ <AODV>: -- J = 0, R = 0, G = 1, D = 1, U = 1
<000000010004> /AODV/ <AODV>: -- hop count = 0, request ID = 0
<000000010004> /AODV/ <AODV>: -- destination = "192.0.0.0", sequence = invalid
<000000010004> /AODV/ <AODV>: -- originator = "192.0.0.4", sequence = 1
<000000010004> /AODV/ <AODV>: ------------------------------------------
<000000010004> /UDP/ <UDP>: creating datagram: source = 654, destination = 654 (32 bytes)
<000000010004> /IPv4/ [IP]: IP-PACKET.request
<000000010004> /IPv4/ <IP>: +++++++++++++++ IP packet +++++++++++++++
<000000010004> /IPv4/ <IP>: ++ version = 4, IHL = 5, type of service = 0, total length = 52
<000000010004> /IPv4/ <IP>: ++ identification = 0000, flags = 0, fragment offset = 0
<000000010004> /IPv4/ <IP>: ++ time to live = 1, protocol = 17, checksum = 0000
<000000010004> /IPv4/ <IP>: ++ source = "192.0.0.4", destination = "255.255.255.255"
<000000010004> /IPv4/ <IP>: +++++++++++++++++++++++++++++++++++++++++
```

This time, the destination address is the limited-broadcast address, and IP does not require a route, since this address maps to the hardware broadcast address of all ones (binary). Hence, the packet is forwarded to the 802.2 LLC, which adds its own header to the frame and passes the resulting PDU to the 802.11 MAC layer for delivery (notice that transmission over the wireless channel has not started yet):

```
<000000010004> /802.2 LLC/ [LLC]: DL-UNITDATA.request: DSAP=aa SSAP=aa Protocol=0000008000
<000000010004> /802.11 MAC/ [MAC]: MA-UNITDATA.request: from 00-00-00-01-00-04 to ff-ff-ff-ff-ff-ff
<000000010004> /802.11 MAC/ MACTX: new tx state = carrierSensePhase2, 1 MPDUs pending
<000000010004> /802.11 MAC/ MACTX: 1 MPDUs in queue. [tx = 2]
```

In addition, AODV creates a new (initially invalid) route table entry for the desired destination, which will be used to track lifetime of the initial ring search with TTL $= 1$ (immediate neighbors only):

```
<000000010004> /AODV/ <AODV>: routing table of "192.0.0.4":
<000000010004> /AODV/ <AODV>: ============================ routing table ============================
<000000010004> /AODV/ <AODV>: == destination      | next hop        | #hops | sequence no. | lifetime | state
<000000010004> /AODV/ <AODV>: == --------------    | --------------- | ----- | ------------ | -------- | --------
<000000010004> /AODV/ <AODV>: ==       192.0.0.0   |        0.0.0.0  |   0   |   invalid    |  240 ms  | requesting
<000000010004> /AODV/ <AODV>: =======================================================================
```

After a DIFS period, during which the channel has been observed to be idle, the medium is seized and the RREQ is actually put "on-the-air":

```
###########################################################
# time = 10.000050,
<000000010004> /802.11 MAC/ MACTX: carrier sense phase 2 timeout - 0 us
<000000010004> /802.11 PHY/ [PHY]: PHY-TXSTART.request
<000000010004> /802.11 MAC/ [MAC]: PHY-TXSTART.confirm
<000000010004> /802.11 MAC/ MPDU: >>>>>>>>>>>>>>> MPDU (data) >>>>>>>>>>>>>>>>
<000000010004> /802.11 MAC/ MPDU: >>> size = 94, frame control = 0008, duration/ID = 0
<000000010004> /802.11 MAC/ MPDU: >>> address1 = ff-ff-ff-ff-ff-ff
<000000010004> /802.11 MAC/ MPDU: >>> address2 = 00-00-00-01-00-04
<000000010004> /802.11 MAC/ MPDU: >>> address3 = 00-00-00-00-00-00
<000000010004> /802.11 MAC/ MPDU: >>> sequence control = 0000
<000000010004> /802.11 MAC/ MPDU: >>> address4 = 00-00-00-00-00-00
<000000010004> /802.11 MAC/ MPDU: >>>>>>>>>>>>>>>>>>>>>>>>>>>>>>>>>>>>>>>>>>>>>
<000000010004> /802.11 PHY/ [PHY]: PHY-DATA.request
<000000010004> /802.11 MAC/ MACTX: new tx state = transmitting, 1 MPDUs pending
```

The four remaining stations are now exposed to a transmission. Anyhow, each station will experience a different propagation delay through the WM, depending on its distance to the transmitter. This is indicated by the respective `time =` prefixes:

```
##############################################################
# time = 10.000050, rxStarted, 0032b9d0, 10.000050, 10.000426

##############################################################
# time = 10.000052, rxStarted, 0032b9d0, 10.000050, 10.000426
<000000010001> /802.11 PHY/ (PHY): WM-RXSTART.indication, 0x0032b9d0, C = 2.84276e-012, TX=0
<000000010001> /802.11 MAC/ (MAC): PHY-CCA.indication (1)
<000000010001> /802.11 MAC/ MACCCA: cca=1, rx=0, tx=0
<000000010001> /802.11 MAC/ (MAC): PHY-RXSTART.indication
<000000010001> /802.11 MAC/ MACRX: new rx state = 1, 0 MPDUs pending
<000000010001> /802.11 PHY/ (PHY): ***********************************************
<000000010001> /802.11 PHY/ (PHY): *** 0x0032b9d0, C = 2.84276e-012, N0 = 4.4043e-014, I = 0, CINR = 18.098627
<000000010001> /802.11 PHY/ (PHY): ***********************************************

##############################################################
# time = 10.000054, rxStarted, 0032b9d0, 10.000050, 10.000426
<000000010002> /802.11 PHY/ (PHY): WM-RXSTART.indication, 0x0032b9d0, C = 3.26114e-013, TX=0
<000000010002> /802.11 MAC/ (MAC): PHY-CCA.indication (1)
<000000010002> /802.11 MAC/ MACCCA: cca=1, rx=0, tx=0
<000000010002> /802.11 PHY/ (PHY): ***********************************************
<000000010002> /802.11 PHY/ (PHY): *** 0x0032b9d0, C = 3.26114e-013, N0 = 4.4043e-014, I = 0, CINR = 8.694929
<000000010002> /802.11 PHY/ (PHY): ***********************************************

##############################################################
# time = 10.000054, rxStarted, 0032b9d0, 10.000050, 10.000426
<000000010003> /802.11 PHY/ (PHY): WM-RXSTART.indication, 0x0032b9d0, C = 2.55268e-013, TX=0
<000000010003> /802.11 MAC/ (MAC): PHY-CCA.indication (1)
<000000010003> /802.11 MAC/ MACCCA: cca=1, rx=0, tx=0
<000000010003> /802.11 PHY/ (PHY): ***********************************************
<000000010003> /802.11 PHY/ (PHY): *** 0x0032b9d0, C = 2.55268e-013, N0 = 4.4043e-014, I = 0, CINR = 7.631199
<000000010003> /802.11 PHY/ (PHY): ***********************************************

##############################################################
# time = 10.000055, rxStarted, 0032b9d0, 10.000050, 10.000426
<000000010000> /802.11 PHY/ (PHY): WM-RXSTART.indication, 0x0032b9d0, C = 1.05467e-013, TX=0
<000000010000> /802.11 MAC/ (MAC): PHY-CCA.indication (1)
<000000010000> /802.11 MAC/ MACCCA: cca=1, rx=0, tx=0
<000000010000> /802.11 PHY/ (PHY): ***********************************************
<000000010000> /802.11 PHY/ (PHY): *** 0x0032b9d0, C = 1.05467e-013, N0 = 4.4043e-014, I = 0, CINR = 3.792403
<000000010000> /802.11 PHY/ (PHY): ***********************************************
```

Notice that station `00-00-00-01-00-04`, i.e. the transmitter, also receives an indication from the WM. However, the 802.11 radio is half-duplex, so this indication is not processed. Other radio models may use this indication to calculate self-interference (for example in FDD systems). Also notice, that only station `00-00-00-01-00-01` is able to demodulate the transmission, due to the 10 dB threshold on the CINR. As a matter of fact, only in this station the MAC is notified of the incoming transmission through a `PHY-RXSTART.indication`. The other MAC instances are only informed about signal energy in the band (physical carrier sense mechanism).

After a certain time (related to the data rate of 2 Mbps), the transmission is completed. Pay attention to the fact that transmission has finished **at the transmitter**, that is, reception is still in progress:

```
##############################################################
# time = 10.000426, txFinished, 0032b9d0, 10.000050, 10.000426
<000000010004> /802.11 PHY/ (PHY): WM-TXEND.indication
<000000010004> /802.11 MAC/ (MAC): PHY-DATA.confirm
<000000010004> /802.11 PHY/ [PHY]: PHY-TXEND.request
<000000010004> /802.11 MAC/ (MAC): PHY-TXEND.confirm
<000000010004> /802.11 MAC/ *MAC*: SSRC or SLRC reset. SSRC=0 SLRC=0 CW=31
<000000010004> /802.2 LLC/ (LLC): MA-UNITDATA-STATUS.indication: SA = "00-00-00-01-00-04", DA = "ff-ff-ff-ff-ff-ff", status = 0
<000000010004> /AODV/ <AODV>: notification: MA-UNITDATA-STATUS.indication ("00-00-00-01-00-04" to "ff-ff-ff-ff-ff-ff", 0)
<000000010004> /802.11 MAC/ MACTX: performing backoff 244 (cca=0, nav=0) [tx = 4]
<000000010004> /802.11 MAC/ MACTX: new tx state = idle, 0 MPDUs pending
```

DCF retry counters and CW are reset because transmission has been successful (remember that multicasts are always counted successful). The back-off procedure is invoked to ensure fairness. Here, an amount of 254 microseconds has been randomly chosen, based on the PHY's slot time and the CW. As the station has no more MPDUs pending, the TX-FSM transitions to its idle state. A short time later, reception has also finished at the sole station that was able to demodulate the transmission. Consequently, the medium is assessed clear again, and the IP packet is delivered to AODV all the way up from PHY, over MAC, LLC, IP, and UDP:

```
#############################################################
# time = 10.000428, rxFinished, 0032b9d0, 10.000050, 10.000426
<000000010001> /802.11 PHY/ (PHY): WM-RXEND.indication, 0x0032b9d0, TRX=0009
<000000010001> /802.11 MAC/ (MAC): PHY-DATA.indication
<000000010001> /802.11 MAC/ (MAC): PHY-RXEND.indication: result = 0
<000000010001> /802.11 MAC/ MACRX: new rx state = 0, 0 MPDUs pending
<000000010001> /802.11 MAC/ MPDU: <<<<<<<<<<<<<<< MPDU (data) <<<<<<<<<<<<<<
<000000010001> /802.11 MAC/ MPDU: <<< size = 94, frame control = 0008, duration/ID = 0
<000000010001> /802.11 MAC/ MPDU: <<< address1 = ff-ff-ff-ff-ff-ff
<000000010001> /802.11 MAC/ MPDU: <<< address2 = 00-00-00-01-00-04
<000000010001> /802.11 MAC/ MPDU: <<< address3 = 00-00-00-00-00-00
<000000010001> /802.11 MAC/ MPDU: <<< sequence control = 0000
<000000010001> /802.11 MAC/ MPDU: <<< address4 = 00-00-00-00-00-00
<000000010001> /802.11 MAC/ MPDU: <<<<<<<<<<<<<<<<<<<<<<<<<<<<<<<<<<<<<<<<
<000000010001> /802.11 MAC/ MACRX: defragmentation of 1 fragments
<000000010001> /802.11 MAC/ MACRX: frame control = 0008, sequence control = 0000, size = 94
<000000010001> /802.11 MAC/ (MAC): PHY-CCA.indication (0)
<000000010001> /802.11 MAC/ MACCCA: cca=0, rx=0, tx=0
<000000010001> /802.2 LLC/ (LLC): MA-UNITDATA.indication: DSAP=aa SSAP=aa Protocol=0000008000
<000000010001> /IPv4/ (IP): DL-UNITDATA.indication
<000000010001> /IPv4/ <IP>: +++++++++++++++ IP packet +++++++++++++++
<000000010001> /IPv4/ <IP>: ++ version = 4, IHL = 5, type of service = 0, total length = 52
<000000010001> /IPv4/ <IP>: ++ identification = 0000, flags = 0, fragment offset = 0
<000000010001> /IPv4/ <IP>: ++ time to live = 1, protocol = 17, checksum = 0000
<000000010001> /IPv4/ <IP>: ++ source = "192.0.0.4", destination = "255.255.255.255"
<000000010001> /IPv4/ <IP>: +++++++++++++++++++++++++++++++++++++++++++++
<000000010001> /UDP/ (UDP): IP-PACKET.indication, 24 bytes from 192.0.0.4:654
<000000010001> /AODV/ (AODV): UDP-DATA.indication
<000000010001> /AODV/ <AODV>: --------------- route request --------------
<000000010001> /AODV/ <AODV>: -- J = 0, R = 0, G = 1, D = 0, U = 1
<000000010001> /AODV/ <AODV>: -- hop count = 0, request ID = 0
<000000010001> /AODV/ <AODV>: -- destination = "192.0.0.0", sequence = invalid
<000000010001> /AODV/ <AODV>: -- originator = "192.0.0.4", sequence = 1
<000000010001> /AODV/ <AODV>: ---------------------------------------------
```

Receiving this RREQ, the intermediate station has learned another – admittedly simple – route to its immediate neighbor: the CBR source. Hence, it adds a new entry to its routing table. Furthermore, it notifies IP about route availability. If there was any pending IP packet waiting for a route to the newly learned destination, this packet would be further processed at this time. In addition AODV receives a notification form the MAC layer that it has received a broadcast frame from station 00-00-00-01-00-04. The WWANS implementation of AODV uses such notifications to monitor the link status to its immediate neighbors:

```
<000000010001> /IPv4/ (IP): route to "192.0.0.4" has become available, next hop "192.0.0.4"
<000000010001> /AODV/ <AODV>: updated route from "192.0.0.1" to "192.0.0.4" (invalid), via "192.0.0.4" (1 hops)
<000000010001> /AODV/ <AODV>: routing table of "192.0.0.1":
<000000010001> /AODV/ <AODV>: ================================ routing table ================================
<000000010001> /AODV/ <AODV>: == destination    | next hop        | #hops | sequence no. | lifetime | state
<000000010001> /AODV/ <AODV>: == ---------------|-----------------|-------|--------------|----------|--------
<000000010001> /AODV/ <AODV>: ==      192.0.0.4 |       192.0.0.4 |     1 |            1 |  5520 ms | active
<000000010001> /AODV/ <AODV>: ===============================================================================
<000000010001> /AODV/ <AODV>: notification: MA-UNITDATA.indication ("00-00-00-01-00-04" to "ff-ff-ff-ff-ff-ff")
```

However, this intermediate station does not have a route for the desired destination, and the TTL field of the RREQ prohibited to forward this RREQ any farther. The CBR traffic source recognizes that it has not received an RREP within the 240 ms timeout it had originally chosen. Hence, it reattempts to find a suitable source by broadcasting another RREQ, this time with an enlarged search range of three hops:

```
#############################################################
# time = 10.240000,
<000000010004> /AODV/ <AODV>: timeout while waiting for RREP ("192.0.0.0"), retry #1
<000000010004> /AODV/ <AODV>: --------------- route request --------------
<000000010004> /AODV/ <AODV>: -- J = 0, R = 0, G = 1, D = 0, U = 1
<000000010004> /AODV/ <AODV>: -- hop count = 0, request ID = 1
<000000010004> /AODV/ <AODV>: -- destination = "192.0.0.0", sequence = invalid
<000000010004> /AODV/ <AODV>: -- originator = "192.0.0.4", sequence = 2
<000000010004> /AODV/ <AODV>: ---------------------------------------------
<000000010004> /UDP/ <UDP>: creating datagram: source = 654, destination = 654 (32 bytes)
<000000010004> /IPv4/ [IP]: IP-PACKET.request
<000000010004> /IPv4/ <IP>: +++++++++++++++ IP packet +++++++++++++++
<000000010004> /IPv4/ <IP>: ++ version = 4, IHL = 5, type of service = 0, total length = 52
<000000010004> /IPv4/ <IP>: ++ identification = 0000, flags = 0, fragment offset = 0
<000000010004> /IPv4/ <IP>: ++ time to live = 3, protocol = 17, checksum = 0000
<000000010004> /IPv4/ <IP>: ++ source = "192.0.0.4", destination = "255.255.255.255"
<000000010004> /IPv4/ <IP>: +++++++++++++++++++++++++++++++++++++++++++++
```

Notice that MAC and PHY layer interactions have been omitted. In addition, the requester accounts for the enlarged search range by waiting for more than the original 240 ms. This time

400 ms are allowed according to the three-hop ring diameter:

```
<000000010004> /AODV/ <AODV>: routing table of "192.0.0.4":
<000000010004> /AODV/ <AODV>: ================================= routing table =================================
<000000010004> /AODV/ <AODV>: == destination  | next hop       | #hops | sequence no. | lifetime | state
<000000010004> /AODV/ <AODV>: == ------------- | -------------- | ----- | ------------ | -------- | --------
<000000010004> /AODV/ <AODV>: ==    192.0.0.0 |       0.0.0.0  |     0 |      invalid |   400 ms | requesting
<000000010004> /AODV/ <AODV>: ================================================================================
```

Again, 00-00-00-01-00-01 is the only station able of successful reception. First, it updates the previously learned route, since the sequence counter is greater than before:

```
<000000010001> /IPv4/ <IP>: +++++++++++++++++ IP packet +++++++++++++++++
<000000010001> /IPv4/ <IP>: ++ version = 4, IHL = 5, type of service = 0, total length = 52
<000000010001> /IPv4/ <IP>: ++ identification = 0000, flags = 0, fragment offset = 0
<000000010001> /IPv4/ <IP>: ++ time to live = 3, protocol = 17, checksum = 0000
<000000010001> /IPv4/ <IP>: ++ source = "192.0.4", destination = "255.255.255.255"
<000000010001> /IPv4/ <IP>: +++++++++++++++++++++++++++++++++++++++++++++
<000000010001> /UDP/ (UDP): IP-PACKET.indication, 24 bytes from 192.0.0.4:654
<000000010001> /AODV/ (AODV): UDP-DATA.indication
<000000010001> /AODV/ <AODV>: --------------- route request --------------
<000000010001> /AODV/ <AODV>: -- J = 0, R = 0, G = 1, D = 0, U = 1
<000000010001> /AODV/ <AODV>: -- hop count = 0, request ID = 1
<000000010001> /AODV/ <AODV>: -- destination = "192.0.0.0", sequence = invalid
<000000010001> /AODV/ <AODV>: -- originator = "192.0.0.4", sequence = 2
<000000010001> /AODV/ <AODV>: -------------------------------------------
<000000010001> /AODV/ <AODV>: updated route from "192.0.0.1" to "192.0.0.4" (2), via "192.0.0.4" (1 hops)
```

It still does not have a valid route, but this time the TTL allows disseminating the RREQ. It increases the hop count, decreases the TTL by one, and re-broadcasts the RREQ:

```
<000000010001> /AODV/ <AODV>: no appropriate route available, forwarding RREQ
<000000010001> /AODV/ <AODV>: --------------- route request --------------
<000000010001> /AODV/ <AODV>: -- J = 0, R = 0, G = 1, D = 0, U = 1
<000000010001> /AODV/ <AODV>: -- hop count = 1, request ID = 1
<000000010001> /AODV/ <AODV>: -- destination = "192.0.0.0", sequence = invalid
<000000010001> /AODV/ <AODV>: -- originator = "192.0.0.4", sequence = 2
<000000010001> /AODV/ <AODV>: -------------------------------------------
<000000010001> /UDP/ <UDP>: creating datagram: source = 654, destination = 654 (32 bytes)
<000000010001> /IPv4/ [IP]: IP-PACKET.request
<000000010001> /IPv4/ <IP>: +++++++++++++++++ IP packet +++++++++++++++++
<000000010001> /IPv4/ <IP>: ++ version = 4, IHL = 5, type of service = 0, total length = 52
<000000010001> /IPv4/ <IP>: ++ identification = 0000, flags = 0, fragment offset = 0
<000000010001> /IPv4/ <IP>: ++ time to live = 2, protocol = 17, checksum = 0000
<000000010001> /IPv4/ <IP>: ++ source = "192.0.0.1", destination = "255.255.255.255"
<000000010001> /IPv4/ <IP>: +++++++++++++++++++++++++++++++++++++++++++++
```

Two stations receive this broadcast: the CBR source, which originated the RREQ, and station 00-00-00-01-00-03, which is situated on the path to 192.0.0.0. The CBR source learns a new route to its neighbor 192.0.0.1, but otherwise blocks processing of its own request, i.e. it does not re-broadcast the RREQ, although the TTL would allow to do so:

```
<000000010004> /IPv4/ <IP>: +++++++++++++++++ IP packet +++++++++++++++++
<000000010004> /IPv4/ <IP>: ++ version = 4, IHL = 5, type of service = 0, total length = 52
<000000010004> /IPv4/ <IP>: ++ identification = 0000, flags = 0, fragment offset = 0
<000000010004> /IPv4/ <IP>: ++ time to live = 2, protocol = 17, checksum = 0000
<000000010004> /IPv4/ <IP>: ++ source = "192.0.0.1", destination = "255.255.255.255"
<000000010004> /IPv4/ <IP>: +++++++++++++++++++++++++++++++++++++++++++++
<000000010004> /UDP/ (UDP): IP-PACKET.indication, 24 bytes from 192.0.0.1:654
<000000010004> /AODV/ (AODV): UDP-DATA.indication
<000000010004> /AODV/ <AODV>: --------------- route request --------------
<000000010004> /AODV/ <AODV>: -- J = 0, R = 0, G = 1, D = 0, U = 1
<000000010004> /AODV/ <AODV>: -- hop count = 1, request ID = 1
<000000010004> /AODV/ <AODV>: -- destination = "192.0.0.0", sequence = invalid
<000000010004> /AODV/ <AODV>: -- originator = "192.0.0.4", sequence = 2
<000000010004> /AODV/ <AODV>: -------------------------------------------
<000000010004> /IPv4/ (IP): route to "192.0.0.1" has become available, next hop "192.0.0.1"
<000000010004> /AODV/ <AODV>: updated route from "192.0.0.4" to "192.0.0.1" (invalid), via "192.0.0.1" (1 hops)
<000000010004> /AODV/ <AODV>: RREQ "192.0.0.4" (1) blocked for 5.6 seconds
```

The other station, 00-00-00-01-00-03, also learns the route to its immediate neighbor. In addition, it knows that this neighbor must be on the path to the originator of the RREQ (the CBR source):

```
<000000010003> /IPv4/ <IP>: +++++++++++++++++ IP packet +++++++++++++++++
<000000010003> /IPv4/ <IP>: ++ version = 4, IHL = 5, type of service = 0, total length = 52
<000000010003> /IPv4/ <IP>: ++ identification = 0000, flags = 0, fragment offset = 0
<000000010003> /IPv4/ <IP>: ++ time to live = 2, protocol = 17, checksum = 0000
<000000010003> /IPv4/ <IP>: ++ source = "192.0.0.1", destination = "255.255.255.255"
```

```
<000000010003> /IPv4/ <IP>: +++++++++++++++++++++++++++++++++++++++++++
<000000010003> /UDP/ (UDP): IP-PACKET.indication, 24 bytes from 192.0.0.1:654
<000000010003> /AODV/ (AODV): UDP-DATA.indication
<000000010003> /AODV/ <AODV>: -------------- route request -------------
<000000010003> /AODV/ <AODV>: -- J = 0, R = 0, G = 1, D = 0, U = 1
<000000010003> /AODV/ <AODV>: -- hop count = 1, request ID = 1
<000000010003> /AODV/ <AODV>: -- destination = "192.0.0.0", sequence = invalid
<000000010003> /AODV/ <AODV>: -- originator = "192.0.0.4", sequence = 2
<000000010003> /AODV/ <AODV>: -----------------------------------------
<000000010003> /IPv4/ (IP): route to "192.0.0.1" has become available, next hop "192.0.0.1"
<000000010003> /AODV/ <AODV>: updated route from "192.0.0.3" to "192.0.0.1" (invalid), via "192.0.0.1" (1 hops)
<000000010003> /IPv4/ (IP): route to "192.0.0.4" has become available, next hop "192.0.0.1"
<000000010003> /AODV/ <AODV>: updated route from "192.0.0.3" to "192.0.0.4" (2), via "192.0.0.1" (2 hops)
<000000010003> /AODV/ <AODV>: routing table of "192.0.0.3":
<000000010003> /AODV/ <AODV>: ============================== routing table ==============================
<000000010003> /AODV/ <AODV>: == destination    | next hop       | #hops | sequence no. | lifetime | state
<000000010003> /AODV/ <AODV>: == -------------- | -------------- | ----- | ------------ | -------- | --------
<000000010003> /AODV/ <AODV>: ==     192.0.0.1  |    192.0.0.1   |   1   |   invalid    | 3000 ms  | active
<000000010003> /AODV/ <AODV>: ==     192.0.0.4  |    192.0.0.1   |   2   |       2      | 5440 ms  | active
<000000010003> /AODV/ <AODV>: ==========================================================================
```

But 00-00-00-01-00-03 does not have a route to the desired destination, either. However, it is allowed to disseminate the RREQ, provided that it increases the hop count field, and decreases the TTL:

```
<000000010003> /AODV/ <AODV>: no appropriate route available, forwarding RREQ
<000000010003> /AODV/ <AODV>: -------------- route request -------------
<000000010003> /AODV/ <AODV>: -- J = 0, R = 0, G = 1, D = 0, U = 1
<000000010003> /AODV/ <AODV>: -- hop count = 2, request ID = 1
<000000010003> /AODV/ <AODV>: -- destination = "192.0.0.0", sequence = invalid
<000000010003> /AODV/ <AODV>: -- originator = "192.0.0.4", sequence = 2
<000000010003> /AODV/ <AODV>: -----------------------------------------
<000000010003> /UDP/ <UDP>: creating datagram: source = 654, destination = 654 (32 bytes)
<000000010003> /IPv4/ [IP]: IP-PACKET.request
<000000010003> /IPv4/ <IP>: +++++++++++++++++ IP packet +++++++++++++++++
<000000010003> /IPv4/ <IP>: ++ version = 4, IHL = 5, type of service = 0, total length = 52
<000000010003> /IPv4/ <IP>: ++ identification = 0000, flags = 0, fragment offset = 0
<000000010003> /IPv4/ <IP>: ++ time to live = 1, protocol = 17, checksum = 0000
<000000010003> /IPv4/ <IP>: ++ source = "192.0.0.3", destination = "255.255.255.255"
<000000010003> /IPv4/ <IP>: +++++++++++++++++++++++++++++++++++++++++++
```

And now, after three hops at $t \approx 10.242$, the RREQ finally reaches its intended destination. Thereby the destination also learns the return path back to the source. This is helpful for TCP, for example, where the destination is required to respond with L4 acknowledgment messages:

```
<000000010000> /IPv4/ <IP>: +++++++++++++++++ IP packet +++++++++++++++++
<000000010000> /IPv4/ <IP>: ++ version = 4, IHL = 5, type of service = 0, total length = 52
<000000010000> /IPv4/ <IP>: ++ identification = 0000, flags = 0, fragment offset = 0
<000000010000> /IPv4/ <IP>: ++ time to live = 1, protocol = 17, checksum = 0000
<000000010000> /IPv4/ <IP>: ++ source = "192.0.0.3", destination = "255.255.255.255"
<000000010000> /IPv4/ <IP>: +++++++++++++++++++++++++++++++++++++++++++
<000000010000> /UDP/ (UDP): IP-PACKET.indication, 24 bytes from 192.0.0.3:654
<000000010000> /AODV/ (AODV): UDP-DATA.indication
<000000010000> /AODV/ <AODV>: -------------- route request -------------
<000000010000> /AODV/ <AODV>: -- J = 0, R = 0, G = 1, D = 0, U = 1
<000000010000> /AODV/ <AODV>: -- hop count = 2, request ID = 1
<000000010000> /AODV/ <AODV>: -- destination = "192.0.0.0", sequence = invalid
<000000010000> /AODV/ <AODV>: -- originator = "192.0.0.4", sequence = 2
<000000010000> /AODV/ <AODV>: -----------------------------------------
<000000010000> /IPv4/ (IP): route to "192.0.0.3" has become available, next hop "192.0.0.3"
<000000010000> /AODV/ <AODV>: updated route from "192.0.0.0" to "192.0.0.3" (invalid), via "192.0.0.3" (1 hops)
<000000010000> /IPv4/ (IP): route to "192.0.0.4" has become available, next hop "192.0.0.3"
<000000010000> /AODV/ <AODV>: updated route from "192.0.0.0" to "192.0.0.4" (2), via "192.0.0.3" (3 hops)
```

Moreover, the destination also recognizes it being the desired endpoint[2]. Thus, it responds with an RREP, which eventually travels along the return path, all the way back to the originator of the RREQ:

```
<000000010000> /AODV/ <AODV>: this is the destination, responding with RREP
<000000010000> /AODV/ <AODV>: -------------- route reply ---------------
<000000010000> /AODV/ <AODV>: -- R = 0, A = 0, prefix size = 0, hop count = 0
<000000010000> /AODV/ <AODV>: -- destination = "192.0.0.0", sequence = 0
<000000010000> /AODV/ <AODV>: -- originator = "192.0.0.4"
<000000010000> /AODV/ <AODV>: -- lifetime = 6000 ms
<000000010000> /AODV/ <AODV>: -----------------------------------------
<000000010000> /UDP/ <UDP>: creating datagram: source = 654, destination = 654 (28 bytes)
<000000010000> /UDP/ [UDP]: forwarding 28 bytes to IP
<000000010000> /IPv4/ [IP]: IP-PACKET.request
```

---

[2]Notice that the procedure would be similar if an intermediate station had a "fresh enough" route to the desired destination. In this case it would send a gratuitous RREP if the according bit flag in the header allowed to do so (G = 1)

```
<000000010000> /IPv4/ <IP>: ++++++++++++++++ IP packet ++++++++++++++++
<000000010000> /IPv4/ <IP>: ++ version = 4, IHL = 5, type of service = 0, total length = 48
<000000010000> /IPv4/ <IP>: ++ identification = 0000, flags = 0, fragment offset = 0
<000000010000> /IPv4/ <IP>: ++ time to live = 1, protocol = 17, checksum = 0000
<000000010000> /IPv4/ <IP>: ++ source = "192.0.0.0", destination = "192.0.0.3"
<000000010000> /IPv4/ <IP>: +++++++++++++++++++++++++++++++++++++++++
```

Now, the ARP protocol comes into play. In contrast to the RREQs, which were destined to the limited broadcast address, the RREP is unicast to its predecessor en route to the source. As a matter of fact, IP needs to have the corresponding hardware address. Hence, IP requests ARP to return the desired hardware address, which it cannot do immediately since no appropriate entry exists in the ARP table. Subsequently, an ARP request is prepared, seeking for the desired hardware address:

```
<000000010000> /ARP/ [ARP]: ARP-LOOKUP.request, not in table
<000000010000> /ARP/ <ARP>: @@@ opcode = request
<000000010000> /ARP/ <ARP>: @@@ sender hardware address = "00-00-00-01-00-00", sender protocol address = "192.0.0.0"
<000000010000> /ARP/ <ARP>: @@@ target hardware address = "00-00-00-00-00-00", target protocol address = "192.0.0.3"
```

The wanted station, 00-00-00-01-00-03, receives this ARP request shortly after, adds the sender's mapping to its local table, and informs IP about the new mapping, which could use it to process packets waiting for an address resolution:

```
<000000010003> /802.2 LLC/ (LLC): MA-UNITDATA.indication: DSAP=aa SSAP=aa Protocol=0000008006
<000000010003> /ARP/ (ARP): DL-UNITDATA.indication
<000000010003> /ARP/ <ARP>: @@@ opcode = request
<000000010003> /ARP/ <ARP>: @@@ sender hardware address = "00-00-00-01-00-00", sender protocol address = "192.0.0.0"
<000000010003> /ARP/ <ARP>: @@@ target hardware address = "00-00-00-00-00-00", target protocol address = "192.0.0.3"
<000000010003> /ARP/ <ARP>: address mapping table of "192.0.0.3":
<000000010003> /ARP/ <ARP>: @@@@@@@@@@@@@@@ arp table @@@@@@@@@@@@@@@@@@
<000000010003> /ARP/ <ARP>: @@ ip address    | hardware address  @@
<000000010003> /ARP/ <ARP>: @@ -------------- | ----------------- @@
<000000010003> /ARP/ <ARP>: @@       192.0.0.0 | 00-00-00-01-00-00 @@
<000000010003> /ARP/ <ARP>: @@@@@@@@@@@@@@@@@@@@@@@@@@@@@@@@@@@@@@@@@@@@@@
<000000010003> /IPv4/ (IP): ARP-LOOKUP.indication ("192.0.0.0" == "00-00-00-01-00-00")
```

Subsequently, it initiates an ARP response bearing the desired mapping:

```
<000000010003> /ARP/ <ARP>: @@@ opcode = reply
<000000010003> /ARP/ <ARP>: @@@ sender hardware address = "00-00-00-01-00-03", sender protocol address = "192.0.0.3"
<000000010003> /ARP/ <ARP>: @@@ target hardware address = "00-00-00-01-00-00", target protocol address = "192.0.0.0"
<000000010003> /802.2 LLC/ [LLC]: DL-UNITDATA.request: DSAP=aa SSAP=aa Protocol=0000008006
```

Station 00-00-00-01-00-00 is waiting for exactly this response, so as to translate 192.0.0.3 into 00-00-00-01-00-03 for a unicast transmission of the RREP.

```
<000000010000> /802.2 LLC/ (LLC): MA-UNITDATA.indication: DSAP=aa SSAP=aa Protocol=0000008006
<000000010000> /ARP/ (ARP): DL-UNITDATA.indication
<000000010000> /ARP/ <ARP>: @@@ opcode = reply
<000000010000> /ARP/ <ARP>: @@@ sender hardware address = "00-00-00-01-00-03", sender protocol address = "192.0.0.3"
<000000010000> /ARP/ <ARP>: @@@ target hardware address = "00-00-00-01-00-00", target protocol address = "192.0.0.0"
<000000010000> /ARP/ <ARP>: address mapping table of "192.0.0.0":
<000000010000> /ARP/ <ARP>: @@@@@@@@@@@@@@@ arp table @@@@@@@@@@@@@@@@@@
<000000010000> /ARP/ <ARP>: @@ ip address    | hardware address  @@
<000000010000> /ARP/ <ARP>: @@ -------------- | ----------------- @@
<000000010000> /ARP/ <ARP>: @@       192.0.0.3 | 00-00-00-01-00-03 @@
<000000010000> /ARP/ <ARP>: @@@@@@@@@@@@@@@@@@@@@@@@@@@@@@@@@@@@@@@@@@@@@@
<000000010000> /IPv4/ (IP): ARP-LOOKUP.indication ("192.0.0.3" == "00-00-00-01-00-03")
<000000010000> /802.2 LLC/ [LLC]: DL-UNITDATA.request: DSAP=aa SSAP=aa Protocol=0000008000
<000000010000> /802.11 MAC/ [MAC]: MA-UNITDATA.request: from 00-00-00-01-00-00 to 00-00-00-01-00-03
```

Station 00-00-00-01-00-03 receives the RREP. Notice, that this is the first unicast transmission received so far. Therefore, for the first time, an L2 ACK frame is transmitted in response, before the MSDU is delivered to the LLC:

```
##############################################################
# time = 10.243640, rxFinished, 00f037f0, 10.243278, 10.243638
<000000010003> /802.11 PHY/ (PHY): WM-RXEND.indication, 0x00f037f0, TRX=0009
<000000010003> /802.11 MAC/ (MAC): PHY-DATA.indication
<000000010003> /802.11 MAC/ (MAC): PHY-RXEND.indication: result = 0
<000000010003> /802.11 MAC/ MACRX: new rx state = 0, 0 MPDUs pending
<000000010003> /802.11 MAC/ MPDU: <<<<<<<<<<<<<< MPDU (data) <<<<<<<<<<<<<<
<000000010003> /802.11 MAC/ MPDU: <<< size = 90, frame control = 0008, duration/ID = 66
<000000010003> /802.11 MAC/ MPDU: <<< address1 = 00-00-00-01-00-03
<000000010003> /802.11 MAC/ MPDU: <<< address2 = 00-00-00-01-00-00
```

```
<000000010003> /802.11 MAC/ MPDU: <<< address3 = 00-00-00-00-00-00
<000000010003> /802.11 MAC/ MPDU: <<< sequence control = 0010
<000000010003> /802.11 MAC/ MPDU: <<< address4 = 00-00-00-00-00-00
<000000010003> /802.11 MAC/ MPDU: <<<<<<<<<<<<<<<<<<<<<<<<<<<<<<<<<<<<<
<000000010003> /802.11 MAC/ MACRX: responding with ACK
<000000010003> /802.11 MAC/ MACTX: new tx state = waitingForSIFS, 1 MPDUs pending
<000000010003> /802.11 MAC/ MACTX: 1 MPDUs in queue. [tx = 7]
<000000010003> /802.11 MAC/ MACRX: new rx state = 2, 0 MPDUs pending
<000000010003> /802.11 MAC/ MACRX: defragmentation of 1 fragments
<000000010003> /802.11 MAC/ MACRX: frame control = 0008, sequence control = 0010, size = 90
```

Next, the station updates its routing table accordingly:

```
<000000010003> /802.2 LLC/ (LLC): MA-UNITDATA.indication: DSAP=aa SSAP=aa Protocol=0000008000
<000000010003> /IPv4/ (IP): DL-UNITDATA.indication
<000000010003> /IPv4/ <IP>: +++++++++++++++ IP packet ++++++++++++++++
<000000010003> /IPv4/ <IP>: ++ version = 4, IHL = 5, type of service = 0, total length = 48
<000000010003> /IPv4/ <IP>: ++ identification = 0000, flags = 0, fragment offset = 0
<000000010003> /IPv4/ <IP>: ++ time to live = 1, protocol = 17, checksum = 0000
<000000010003> /IPv4/ <IP>: ++ source = "192.0.0.0", destination = "192.0.0.3"
<000000010003> /IPv4/ <IP>: +++++++++++++++++++++++++++++++++++++++++++
<000000010003> /UDP/ (UDP): IP-PACKET.indication, 20 bytes from 192.0.0.0:654
<000000010003> /AODV/ (AODV): UDP-DATA.indication
<000000010003> /AODV/ <AODV>: --------------- route reply ---------------
<000000010003> /AODV/ <AODV>: -- R = 0, A = 0, prefix size = 0, hop count = 0
<000000010003> /AODV/ <AODV>: -- destination = "192.0.0.0", sequence = 0
<000000010003> /AODV/ <AODV>: -- originator = "192.0.0.4"
<000000010003> /AODV/ <AODV>: -- lifetime = 6000 ms
<000000010003> /AODV/ <AODV>: ---------------------------------------
<000000010003> /IPv4/ (IP): route to "192.0.0.0" has become available, next hop "192.0.0.0"
<000000010003> /AODV/ <AODV>: updated route from "192.0.0.3" to "192.0.0.0" (invalid), via "192.0.0.0" (1 hops)
<000000010003> /AODV/ <AODV>: routing table of "192.0.0.3":
<000000010003> /AODV/ <AODV>: =============================== routing table ===============================
<000000010003> /AODV/ <AODV>: == destination | next hop |  #hops | sequence no. | lifetime | state
<000000010003> /AODV/ <AODV>: == --------------- | -------------- | ----- | ------------ | -------- | --------
<000000010003> /AODV/ <AODV>: ==       192.0.0.0 |      192.0.0.0 |    1 |            0 | 6000 ms  | active
<000000010003> /AODV/ <AODV>: ==       192.0.0.1 |      192.0.0.1 |    1 |      invalid | 3000 ms  | active
<000000010003> /AODV/ <AODV>: ==       192.0.0.4 |      192.0.0.1 |    2 |            2 | 5437 ms  | active
<000000010003> /AODV/ <AODV>: =============================================================================
```

Then, the RREP is queued to be forwarded further along the path to the RREQ's originator. Notice that the ARP request/response sequence necessary to determine the hardware address of 192.0.0.1 has been omitted in the following:

```
<000000010003> /AODV/ <AODV>: forwarding RREP towards originator
<000000010003> /AODV/ <AODV>: --------------- route reply ---------------
<000000010003> /AODV/ <AODV>: -- R = 0, A = 0, prefix size = 0, hop count = 1
<000000010003> /AODV/ <AODV>: -- destination = "192.0.0.0", sequence = 0
<000000010003> /AODV/ <AODV>: -- originator = "192.0.0.4"
<000000010003> /AODV/ <AODV>: -- lifetime = 6000 ms
<000000010003> /AODV/ <AODV>: ---------------------------------------
<000000010003> /UDP/ <UDP>: creating datagram: source = 654, destination = 654 (28 bytes)
<000000010003> /UDP/ [UDP]: forwarding 28 bytes to IP
<000000010003> /IPv4/ [IP]: IP-PACKET.request
<000000010003> /IPv4/ <IP>: +++++++++++++++ IP packet ++++++++++++++++
<000000010003> /IPv4/ <IP>: ++ version = 4, IHL = 5, type of service = 0, total length = 48
<000000010003> /IPv4/ <IP>: ++ identification = 0000, flags = 0, fragment offset = 0
<000000010003> /IPv4/ <IP>: ++ time to live = 1, protocol = 17, checksum = 0000
<000000010003> /IPv4/ <IP>: ++ source = "192.0.0.3", destination = "192.0.0.1"
<000000010003> /IPv4/ <IP>: +++++++++++++++++++++++++++++++++++++++++++
<000000010003> /AODV/ [AODV]: route request
<000000010003> /IPv4/ (IP): route to "192.0.0.1" has become available, next hop "192.0.0.1"
<000000010003> /AODV/ <AODV>: maintaining lifetimes for active routes
```

In accordance to DCF basic access rules, the ACK frame is transmitted after a SIFS period, without any carrier sense at all. It can also be seen that two MPDUs are queued for transmission, namely the ACK frame and the ARP request:

```
############################################################
# time = 10.243650,
<000000010003> /802.11 MAC/ MACTX: SIFS has expired, continuing transmission
<000000010003> /802.11 PHY/ [PHY]: PHY-TXSTART.request
<000000010003> /802.11 MAC/ (MAC): PHY-TXSTART.confirm
<000000010003> /802.11 MAC/ MPDU: >>>>>>>>>>>>>>> MPDU (+ack) >>>>>>>>>>>>>>>
<000000010003> /802.11 MAC/ MPDU: >>> size = 18, frame control = 00d4, duration/ID = 0
<000000010003> /802.11 MAC/ MPDU: >>> address1 = 00-00-00-01-00-00
<000000010003> /802.11 MAC/ MPDU: >>>>>>>>>>>>>>>>>>>>>>>>>>>>>>>>>>>>>
<000000010003> /802.11 PHY/ [PHY]: PHY-DATA.request
<000000010003> /802.11 MAC/ MACTX: new tx state = transmitting, 2 MPDUs pending
```

After a short period, the ACK frame has been received, and the MPDU is dropped from the TX queue. The back-off procedure is performed, and the station chooses a random delay of 307 microseconds.

```
##############################################################
# time = 10.243724, rxFinished, 00f037f0, 10.243650, 10.243722
<000000010000> /802.11 PHY/ (PHY): WM-RXEND.indication, 0x00f037f0, TRX=0009
<000000010000> /802.11 MAC/ (MAC): PHY-DATA.indication
<000000010000> /802.11 MAC/ (MAC): PHY-RXEND.indication: result = 0
<000000010000> /802.11 MAC/ MACRX: new rx state = 0, 0 MPDUs pending
<000000010000> /802.11 MAC/ MPDU: <<<<<<<<<<<<< MPDU (+ack) <<<<<<<<<<<<<
<000000010000> /802.11 MAC/ MPDU: <<< size = 18, frame control = 00d4, duration/ID = 0
<000000010000> /802.11 MAC/ MPDU: <<< address1 = 00-00-00-01-00-00
<000000010000> /802.11 MAC/ MPDU: <<<<<<<<<<<<<<<<<<<<<<<<<<<<<<
<000000010000> /802.11 MAC/ MACTX: received positive ACK, removing frame from TX queue
<000000010000> /802.11 MAC/ *MAC*: SSRC or SLRC reset. SSRC=0 SLRC=0 CW=31
<000000010000> /802.2 LLC/ (LLC): MA-UNITDATA-STATUS.indication: SA = "00-00-00-01-00-00", DA = "00-00-00-01-00-03", status = 0
<000000010000> /AODV/ <AODV>: notification: MA-UNITDATA-STATUS.indication ("00-00-00-01-00-00" to "00-00-00-01-00-03", 0)
<000000010000> /802.11 MAC/ MACTX: performing backoff 307 (cca=1, nav=0) [tx = 6]
<000000010000> /802.11 MAC/ MACTX: new tx state = idle, 0 MPDUs pending
<000000010000> /802.11 MAC/ (MAC): PHY-CCA.indication (0)
<000000010000> /802.11 MAC/ MACCCA: cca=0, rx=0, tx=0
```

In the end, after more ARP requests and replies, ACKs and RREP forwarding, the RREP finally reaches the originator of the corresponding RREQ. At this time – approximately 250 ms after CBR has pushed the first user data onto the IP queue – the CBR source knows, to which station the user data actually needs to be sent:

```
##############################################################
# time = 10.247743, rxFinished, 00f06358, 10.247381, 10.247741
<000000010004> /802.11 PHY/ (PHY): WM-RXEND.indication, 0x00f06358, TRX=0009
<000000010004> /802.11 MAC/ (MAC): PHY-DATA.indication
<000000010004> /802.11 MAC/ (MAC): PHY-RXEND.indication: result = 0
<000000010004> /802.11 MAC/ MACRX: new rx state = 0, 0 MPDUs pending
<000000010004> /802.11 MAC/ MPDU: <<<<<<<<<<<<< MPDU (data) <<<<<<<<<<<<<
<000000010004> /802.11 MAC/ MPDU: <<< size = 90, frame control = 0008, duration/ID = 66
<000000010004> /802.11 MAC/ MPDU: <<< address1 = 00-00-00-01-00-04
<000000010004> /802.11 MAC/ MPDU: <<< address2 = 00-00-00-01-00-01
<000000010004> /802.11 MAC/ MPDU: <<< address3 = 00-00-00-00-00-00
<000000010004> /802.11 MAC/ MPDU: <<< sequence control = 0030
<000000010004> /802.11 MAC/ MPDU: <<< address4 = 00-00-00-00-00-00
<000000010004> /802.11 MAC/ MPDU: <<<<<<<<<<<<<<<<<<<<<<<<<<<<<<<<<<<<
<000000010004> /802.11 MAC/ MACRX: responding with ACK
<000000010004> /802.11 MAC/ MACTX: new tx state = waitingForSIFS, 1 MPDUs pending
<000000010004> /802.11 MAC/ MACTX: 1 MPDUs in queue. [tx = 7]
<000000010004> /802.11 MAC/ MACRX: new rx state = 2, 0 MPDUs pending
<000000010004> /802.11 MAC/ MACRX: defragmentation of 1 fragments
<000000010004> /802.11 MAC/ MACRX: frame control = 0008, sequence control = 0030, size = 90
<000000010004> /802.2 LLC/ (LLC): MA-UNITDATA.indication: DSAP=aa SSAP=aa Protocol=0000008000
<000000010004> /IPv4/ (IP): DL-UNITDATA.indication
<000000010004> /IPv4/ <IP>: ++++++++++++++++ IP packet ++++++++++++++++
<000000010004> /IPv4/ <IP>: ++ version = 4, IHL = 5, type of service = 0, total length = 48
<000000010004> /IPv4/ <IP>: ++ identification = 0000, flags = 0, fragment offset = 0
<000000010004> /IPv4/ <IP>: ++ time to live = 1, protocol = 17, checksum = 0000
<000000010004> /IPv4/ <IP>: ++ source = "192.0.0.1", destination = "192.0.0.4"
<000000010004> /IPv4/ <IP>: +++++++++++++++++++++++++++++++++++++++++
<000000010004> /UDP/ (UDP): IP-PACKET.indication, 20 bytes from 192.0.0.1:654
<000000010004> /AODV/ (AODV): UDP-DATA.indication
<000000010004> /AODV/ <AODV>: --------------- route reply ---------------
<000000010004> /AODV/ <AODV>: -- R = 0, A = 0, prefix size = 0, hop count = 2
<000000010004> /AODV/ <AODV>: -- destination = "192.0.0.0", sequence = 0
<000000010004> /AODV/ <AODV>: -- originator = "192.0.0.4"
<000000010004> /AODV/ <AODV>: -- lifetime = 6000 ms
<000000010004> /AODV/ <AODV>: -------------------------------------------
<000000010004> /AODV/ <AODV>: updated route from "192.0.0.4" to "192.0.0.1" (invalid), via "192.0.0.1" (1 hops)
<000000010004> /AODV/ <AODV>: maintaining lifetimes for active routes
<000000010004> /AODV/ <AODV>: routing table of "192.0.0.4":
<000000010004> /AODV/ <AODV>: ============================= routing table =============================
<000000010004> /AODV/ <AODV>: == destination    | next hop        | #hops | sequence no. | lifetime | state
<000000010004> /AODV/ <AODV>: == --------------- | --------------- | ----- | ------------ | -------- | --------
<000000010004> /AODV/ <AODV>: ==     192.0.0.0   |     192.0.0.1   |   3   |           0  | 6000 ms  | active
<000000010004> /AODV/ <AODV>: ==     192.0.0.1   |     192.0.0.1   |   1   |     invalid  | 3000 ms  | active
<000000010004> /AODV/ <AODV>: =========================================================================
```

Now, it is possible for the CBR source, 192.0.0.4 to transmit the packets intended for the destination, 192.0.0.0 taking the intermediate hops 192.0.0.1, and 192.0.0.3. Every time a route is used to forward a packet, route lifetime is prolonged. When the lifetime has expired, the route becomes inactive at first, but remains in the routing table for a certain time span. If it is not needed within this span, it is completely removed from the table. In the end, the first user data packet created by the CBR source finally arrives at the UDP sink application, about 262 ms after it has been requested for transmission:

```
##############################################################
# time = 10.262287, rxFinished, 00f06358, 10.257909, 10.262285
<000000010000> /802.11 PHY/ (PHY): WM-RXEND.indication, 0x00f06358, TRX=0009
<000000010000> /802.11 MAC/ (MAC): PHY-DATA.indication
<000000010000> /802.11 MAC/ (MAC): PHY-RXEND.indication: result = 0
```

```
<000000010000> /802.11 MAC/ MACRX: new rx state = 0, 0 MPDUs pending
<000000010000> /802.11 MAC/ MPDU: <<<<<<<<<<<<<<< MPDU (data) <<<<<<<<<<<<<<<
<000000010000> /802.11 MAC/ MPDU: <<< size = 1094, frame control = 0008, duration/ID = 66
<000000010000> /802.11 MAC/ MPDU: <<< address1 = 00-00-00-01-00-00
<000000010000> /802.11 MAC/ MPDU: <<< address2 = 00-00-00-01-00-03
<000000010000> /802.11 MAC/ MPDU: <<< address3 = 00-00-00-00-00-00
<000000010000> /802.11 MAC/ MPDU: <<< sequence control = 0040
<000000010000> /802.11 MAC/ MPDU: <<< address4 = 00-00-00-00-00-00
<000000010000> /802.11 MAC/ MPDU: <<<<<<<<<<<<<<<<<<<<<<<<<<<<<<<<<<<<<<<
<000000010000> /802.11 MAC/ MACRX: responding with ACK
<000000010000> /802.11 MAC/ MACTX: new tx state = waitingForSIFS, 1 MPDUs pending
<000000010000> /802.11 MAC/ MACTX: 1 MPDUs in queue. [tx = 7]
<000000010000> /802.11 MAC/ MACRX: new rx state = 2, 0 MPDUs pending
<000000010000> /802.11 MAC/ MACRX: defragmentation of 1 fragments
<000000010000> /802.11 MAC/ MACRX: frame control = 0008, sequence control = 0040, size = 1094
<000000010000> /802.2 LLC/ (LLC): MA-UNITDATA.indication: DSAP=aa SSAP=aa Protocol=0000008000
<000000010000> /IPv4/ (IP): DL-UNITDATA.indication
<000000010000> /IPv4/ <IP>: ++++++++++++++++ IP packet ++++++++++++++++
<000000010000> /IPv4/ <IP>: ++ version = 4, IHL = 5, type of service = 0, total length = 1052
<000000010000> /IPv4/ <IP>: ++ identification = 0000, flags = 0, fragment offset = 0
<000000010000> /IPv4/ <IP>: ++ time to live = 253, protocol = 17, checksum = 0000
<000000010000> /IPv4/ <IP>: ++ source = "192.0.0.4", destination = "192.0.0.0"
<000000010000> /IPv4/ <IP>: +++++++++++++++++++++++++++++++++++++++++++
<000000010000> /UDP/ (UDP): IP-PACKET.indication, 1024 bytes from 192.0.0.4:8100
<000000010000> /UDP sink/ (UDP-SINK): UDP-DATA.indication (1024 bytes)
<000000010000> /AODV/ <AODV>: notification: MA-UNITDATA.indication ("00-00-00-01-00-03" to "00-00-00-01-00-00")
<000000010000> /802.11 MAC/ (MAC): PHY-CCA.indication (0)
<000000010000> /802.11 MAC/ MACCCA: cca=0, rx=2, tx=7
```

Naturally, route discovery and address resolution are not required for the remaining packets. Let us observe when the other packets arrive: 10.514140, 11.013891, 11.514723, 12.014443, ..., 19.514212, 20.014825. Obviously, latency is now much less than the initial 260 ms: approximately 14 ms are required for subsequent packets at a data rate of 2 Mbps.

7.2.5 Conclusion

This first flow was intended to provide some insight into the complex interactions of different simulation models and protocols across all seven layers of the protocol stack. Even this very simple example with only five stations and a single traffic flow reveals the new simulator's capabilities. It should be noted that simulator functions that have not come to use in this first example, e.g. multiple concurrent transmissions, suspending back-off counters, RTS/CTS operation, fragmentation, L2 retries, gratuitous RREPs, etc. have nonetheless been thoroughly tested in countless simulation runs. For obvious reasons, it is not possible to cover all aspects and unique features worth describing.

7.3 Second Example: A TBSD-RRM Flow

The second example is going be presented in less detail than the previous one. After a short overview of tools dedicated to TBSD-RRM simulation flows, some simulator output is discussed, so as to explain how WWANS has been successfully used for the development of this innovative management protocol.

7.3.1 Automatically Generating Link Activation Requests

To WWANS, the //ddcagen(12.5, 0.5) line in the configuration script (listing 7.1) is just a comment line and is ignored in simulations. But actually, this comment is a special directive that is used to control a tool dedicated to the automatic generation of link activation requests, as they are required by TBSD-RRM. This special tool, ddcagen, uses the the original configuration script as a template, and replaces the special comment line by channel acquisition requests.

Remember that the "heterogeneous100" topology presented in chapter 4 had more than 500 edges. It would be impractical to create a request for each of these edges manually. Therefore, ddcagen uses connectivity information to create acquisition requests for all existing links. This connectivity information may be provided in form of a single connectivity graph, as determined by wwans using the "-g" option, or by a number of local NDP graphs. For the previous example, this would look like:

```
// events generated by ddcagen
rrm_acquire("00-00-00-01-00-02", "00-00-00-01-00-03", 25);
rrm_acquire("00-00-00-01-00-04", "00-00-00-01-00-01", 25);
rrm_acquire("00-00-00-01-00-03", "00-00-00-01-00-01", 25);
rrm_acquire("00-00-00-01-00-03", "00-00-00-01-00-00", 25);
rrm_acquire("00-00-00-01-00-00", "00-00-00-01-00-03", 25);
rrm_acquire("00-00-00-01-00-02", "00-00-00-01-00-00", 25);
rrm_acquire("00-00-00-01-00-01", "00-00-00-01-00-04", 25);
rrm_acquire("00-00-00-01-00-00", "00-00-00-01-00-02", 25);
rrm_acquire("00-00-00-01-00-01", "00-00-00-01-00-03", 25);
rrm_acquire("00-00-00-01-00-03", "00-00-00-01-00-02", 25);
```

Here, all link activations are requested simultaneously, but it is also possible to have the requests separated by a time gap, for example 0.5 seconds. This explains the two parameters in the special directive: The first one specifies the time for the first activation, while the second one determines the inter-request gap. In other words, the previous block has been generated out of the following line:

```
//{{ddcagen(25.0, 0)}}}
```

7.3.2 Obtaining Assignment Results

Now, WWANS is used to simulate the configuration script created with the help of ddcagen. Again, debug and release versions can be used for simulation. This time the release version creates a useful output, namely a number of local graphs created with the __rrm_storegraph directive, which has been introduced in the previous chapter. Recall that these graphs represent local TBSD-RRM tables. In addition to these graphs, the debug version is also able to produce verbose output. Two representative portions of this output are shown in listings 7.4 and 7.5. While this is only a fraction of 14 MB of output, it still gives an impression of "under the hood" activity. These portions are the outcome of a simulation run using the "chain10" topology, which took only eight seconds to complete on the mentioned hardware configuration. However, these portions are not going to be commented, due to space limitations. Another interesting portion from this simulation is presented below. Here, multiple (three in this case) concurrent transmissions interfere, while the receiving station is waiting for an L2 ACK response:

```
###############################################################
# time = 25.001516, rxStarted, 00f18400, 25.001514, 25.001586
<000000010005> /802.11 PHY/ (PHY): WM-RXSTART.indication, 0x00f18400, C = 1.6e-012, TX=0
<000000010005> /802.11 PHY/ (PHY): *** 0x00f1d448 carrier lost during reception!
<000000010005> /802.11 MAC/ (MAC): PHY-RXEND.indication: result = 2
<000000010005> /802.11 MAC/ MACRX: new rx state = 0, 0 MPDUs pending
<000000010005> /802.11 MAC/ MACTX: received an invalid frame while waiting for ACK
<000000010005> /802.11 MAC/ *MAC*: MSDURC=1, incremented SSRC. SSRC=1 SLRC=0 CW=63
<000000010005> /802.11 MAC/ MACTX: retransmitting, retry #1, SSRC=1, SLRC=0
<000000010005> /802.11 MAC/ MACTX: performing backoff 1068 (cca=1, nav=0) [tx = 6]
<000000010005> /802.11 MAC/ MACTX: new tx state = carrierSensePhase1, 1 MPDUs pending
<000000010005> /802.11 PHY/ (PHY): *****************************************************
<000000010005> /802.11 PHY/ (PHY): *** 0x00f18400, C = 1.6e-012, N0 = 4.4043e-014, I = 1.60256e-012, CINR = -0.124689
<000000010005> /802.11 PHY/ (PHY): *** 0x00f1b958, C = 2.56e-015, N0 = 4.4043e-014, I = 3.2e-012, CINR = -31.028466
<000000010005> /802.11 PHY/ (PHY): *** 0x00f1d448, C = 1.6e-012, N0 = 4.4043e-014, I = 1.60256e-012, CINR = -0.124689
<000000010005> /802.11 PHY/ (PHY): *****************************************************
```

7.3.3 Evaluating and Verifying Assignment Results

Assume that simulation has finished and a number of local RRM graphs have been created by WWANS. The problem is now: how to say if assignments are at all valid in the global perspective?

```
########################################
# time = 31.017272, rxFinished, 00f11e38, 31.016982, 31.017270
<00000010005> /802.11 PHY    (PHY) WM-RXEND.indication, 0x00f11e38, TRX=0009
<00000010005> /802.11 PHY    (PHY) PHY-DATA.indication
<00000010005> /802.11 MAC:   (MAC) PHY-RXEND.indication: result = 0
<00000010005> /802.11 MAC:   MACRX: new rx state = 0, 0 MPDUs pending
<00000010005> /802.11 MPDU:  <<< MPDU (mgmt) <<<<<<<<<<<<<<<
<00000010005> /802.11 MPDU:  <<< size = 72, frame control = 00d0, duration/ID = 66
<00000010005> /802.11 MPDU:  <<< address1 = 00-00-00-01-00-05
<00000010005> /802.11 MPDU:  <<< address2 = 00-00-00-01-00-06
<00000010005> /802.11 MPDU:  <<< address3 = 00-00-00-00-00-00
<00000010005> /802.11 MPDU:  <<< sequence control = 0220
<00000010005> /802.11 MPDU:  <<<<<<<<<<<<<<<<<<<<<<<<<<<<<<
<00000010005> /802.11 MAC:   MACCCA: cca=0, fx=2, tx=7
<00000010005> /802.11 MAC:   MACRX: defragmentation of 1 fragments
<00000010005> /802.11 MAC:   MACRX: frame control = 00d0, sequence control = 0220, size = 72
<00000010005> /802.11 MAC:   MACTX: responding with ACK
<00000010005> /802.11 MAC:   MACTX: new tx state = waitingForSIFS, 1 MPDUs pending
<00000010005> /802.11 MAC:   MACTX: 1 MPDUs in queue, [tx = 7]
<00000010005> /802.11 MAC:   MACRX: new rx state = 2, 0 MPDUs pending
<00000010005> /802.11 MAC:   (MAC) PHY-CCA.indication (0)
<00000010005> /802.11 MAC/  (MLME): MLME-MANAGEMENTDATA.indication
<00000010005> /RRM:         (RRM): MLME-MANAGEMENTDATA.indication
<00000010005> /RRM/ <RRMV>: received positive vote for 00-00-00-01-00-05 -> 00-00-00-01-00-04 (1)
<00000010005> /RRM/ <RRMV>: incorporating positive vote information
<00000010005> /RRM/ <RRM>:
<00000010005> /RRM/ <RRM>:  resource management table ( 1 entries )
<00000010005> /RRM/ <RRM>:  ----------+-------+--------+-------------------+-------
<00000010005> /RRM/ <RRM>:  station   | state | degree | assigned by       | color
<00000010005> /RRM/ <RRM>:  ----------+-------+--------+-------------------+-------
<00000010005> /RRM/ <RRM>:  station 1 |   2   |   2    | 00-00-00-01-00-05 |   1
<00000010005> /RRM/ <RRM>:  station 2 |   2   |   2    | 00-00-00-01-00-04 |   2
<00000010005> /RRM/ <RRM>:  ----------+-------+--------+-------------------+-------
<00000010005> /RRM/ <RRM>:
<00000010005> /RRM/ <RRM>:  assignments by priority ( 12 entries )
<00000010005> /RRM/ <RRM>:  -------------------+-------+--------+-------------------+-------
<00000010005> /RRM/ <RRM>:  state              | state | degree | assigned by       | color
<00000010005> /RRM/ <RRM>:  -------------------+-------+--------+-------------------+-------
<00000010005> /RRM/ <RRM>:  00-00-00-01-00-08  |   5   |   2    | 00-00-00-01-00-08 |   0
<00000010005> /RRM/ <RRM>:  00-00-00-01-00-07  |   5   |   2    | 00-00-00-01-00-07 |   0
<00000010005> /RRM/ <RRM>:  00-00-00-01-00-06  |   5   |   2    | 00-00-00-01-00-06 |   0
<00000010005> /RRM/ <RRM>:  00-00-00-01-00-05  |   5   |   2    | 00-00-00-01-00-05 |   4
<00000010005> /RRM/ <RRM>:  00-00-00-01-00-04  |   5   |   2    | 00-00-00-01-00-04 |   3
<00000010005> /RRM/ <RRM>:  00-00-00-01-00-07  |   2   |   2    | 00-00-00-01-00-07 |   5
<00000010005> /RRM/ <RRM>:  00-00-00-01-00-06  |   2   |   2    | 00-00-00-01-00-06 |   6
<00000010005> /RRM/ <RRM>:  00-00-00-01-00-05  |   2   |   2    | 00-00-00-01-00-05 |   3
<00000010005> /RRM/ <RRM>:  00-00-00-01-00-04  |   2   |   2    | 00-00-00-01-00-04 |   2
<00000010005> /RRM/ <RRM>:  00-00-00-01-00-03  |   2   |   1    | 00-00-00-01-00-03 |   2
<00000010005> /RRM/ <RRM>:  00-00-00-01-00-04  |   2   |   1    | 00-00-00-01-00-04 |   1
<00000010005> /RRM/ <RRM>:  00-00-00-01-00-03  |   0   |        | 00-00-00-01-00-05 |  -1
<00000010005> /RRM/ <RRM>:  -------------------+-------+--------+-------------------+-------
<00000010005> /RRM/ <RRM>:
<00000010005> /RRM/ <RRMT>:  pending transaction table ( 2 entries )
<00000010005> /RRM/ <RRMT>:  ---------------+-------------------+-------+---------+--------+---------+--------
<00000010005> /RRM/ <RRMT>:  state          | assigned by       | color | sequence| attempt| votes(+)| votes(-)| lifetime
<00000010005> /RRM/ <RRMT>:  ---------------+-------------------+-------+---------+--------+---------+--------
<00000010005> /RRM/ <RRMT>:  collecting     | 00-00-00-01-00-05 |   2   |    2    |   1    |    2    |    1    | 3.992740
<00000010005> /RRM/ <RRMT>:  pending..:     |                   |       |         |        |         |         |
<00000010005> /RRM/ <RRMT>:  vote(+)        |                   |       |         |        |         |         |
<00000010005> /RRM/ <RRMT>:  collecting     | 00-00-00-01-00-06 |   2   |    1    |   0    |    1    |    0    | 3.992740
<00000010005> /RRM/ <RRMT>:  pending..:     |                   |       |         |        |         |         |
<00000010005> /RRM/ <RRMT>:  vote(+)        |                   |       |         |        |         |         |
<00000010005> /RRM/ <RRMT>:  ---------------+-------------------+-------+---------+--------+---------+--------
<00000010005> /RRM/ <RRM>:  graph integrity: ok.
```

Listing 7.4: A Positive Vote is Received. Also Shows RRM and Transaction Table Structures

```
##########################################################
# time = 25.000641, rxFinished, 00f1c670, 25.000062, 25.000638
# (RRM/ (RRM): MLME-MANAGEMENTDATA.indication
<0000000100006> /RRM/ <RRM>:  ----- resource management table ( 3 entries) -----
<0000000100006> /RRM/ <RRM>:   station 1      | station 2      | state | degree | assigned by    | color | sequence
<0000000100006> /RRM/ <RRM>:  +---------------+----------------+-------+--------+----------------+-------+---------
<0000000100006> /RRM/ <RRMS>: 00-00-01-00-04 | ff-ff-ff-ff-ff |   0   |   0    | ff-ff-ff-ff-ff |  -1   |   0
<0000000100006> /RRM/ <RRMS>: 00-00-01-00-05 | 00-00-01-00-04 |   2   |   2    | 00-00-01-00-05 |   3   |   0
<0000000100006> /RRM/ <RRMS>: 00-00-01-00-06 | ff-ff-ff-ff-ff |   0   |   0    | ff-ff-ff-ff-ff |  -1   |   0
<0000000100006> /RRM/ <RRMV>: received a vote query for 00-00-01-00-05 -> 00-00-01-00-04 (3), invited
<0000000100006> /RRM/ <RRM>:  ----- assignments by priority ( 13 entries) -----
<0000000100006> /RRM/ <RRM>:   station 1      | station 2      | state | degree | assigned by    | color | sequence
<0000000100006> /RRM/ <RRM>:  +---------------+----------------+-------+--------+----------------+-------+---------
<0000000100006> /RRM/ <RRMS>: 00-00-01-00-06 | 00-00-01-00-07 |   2   |   2    | 00-00-01-00-06 |   4   |   0
<0000000100006> /RRM/ <RRMS>: 00-00-01-00-06 | 00-00-01-00-05 |   2   |   2    | 00-00-01-00-06 |   3   |   0
<0000000100006> /RRM/ <RRMS>: 00-00-01-00-06 | 00-00-01-00-04 |   2   |   2    | 00-00-01-00-05 |  -1   |   0
<0000000100006> /RRM/ <RRMS>: 00-00-01-00-09 | 00-00-01-00-08 |   0   |   0    | 00-00-01-00-06 |  -1   |   0
<0000000100006> /RRM/ <RRMS>: 00-00-01-00-08 | 00-00-01-00-09 |   0   |   0    | 00-00-01-00-06 |  -1   |   0
<0000000100006> /RRM/ <RRMS>: 00-00-01-00-08 | 00-00-01-00-07 |   0   |   0    | 00-00-01-00-06 |  -1   |   0
<0000000100006> /RRM/ <RRMS>: 00-00-01-00-07 | 00-00-01-00-06 |   0   |   0    | 00-00-01-00-06 |  -1   |   0
<0000000100006> /RRM/ <RRMS>: 00-00-01-00-05 | 00-00-01-00-05 |   0   |   0    | 00-00-01-00-06 |  -1   |   0
<0000000100006> /RRM/ <RRMS>: 00-00-01-00-04 | 00-00-01-00-04 |   0   |   0    | 00-00-01-00-06 |  -1   |   0
<0000000100006> /RRM/ <RRMS>: 00-00-01-00-05 | 00-00-01-00-05 |   0   |   0    | 00-00-01-00-06 |  -1   |   0
<0000000100006> /RRM/ <RRMS>: 00-00-01-00-04 | 00-00-01-00-03 |   0   |   0    | 00-00-01-00-06 |  -1   |   0
<0000000100006> /RRM/ <RRMS>: 00-00-01-00-03 | 00-00-01-00-04 |   0   |   0    | 00-00-01-00-06 |  -1   |   0
<0000000100006> /RRM/ <RRM>:  +---------------+----------------+-------+--------+----------------+-------+---------
<0000000100006> /RRM/ <RRM>:  || 00-00-01-00-05 | 00-00-01-00-04 |   2   |   2    | 00-00-01-00-05 |   3   |   0 |!!
<0000000100006> /RRM/ <RRMV>: graph integrity: ok.
<0000000100006> /RRM/ <RRMV>: channel is not available (1 reasons)
<0000000100006> /RRM/ <RRMV>: casting an immediate negative vote
<0000000100006> /RRM/ <RRMQ>: queueing packet (2 pending)
<0000000100006> /RRM/ <RRMQ>: queue at this time:
<0000000100006> /RRM/ <RRMQ>: packet to "00-00-00-01-00-07", priority = 127, type = query, 3 entries
<0000000100006> /RRM/ <RRMQ>: packet to "00-00-00-01-00-05", priority = 23, type = vote(-), 2 entries
<0000000100006> /RRM/ <RRMQ>: packet to "00-00-00-01-00-07", priority = 0, type = query, 3 entries
<0000000100006> /RRM/ <RRMT>: ----- pending transaction table ( 3 entries) -----
<0000000100006> /RRM/ <RRMT>:   station 1         | station 2         | color | sequence | state      | assigned by       | attempt | votes(+) | votes(-) | lifetime
<0000000100006> /RRM/ <RRMT>:  +------------------+-------------------+-------+----------+------------+-------------------+---------+----------+----------+---------
<0000000100006> /RRM/ <RRMT>: 00-00-00-01-00-04 | 00-00-00-01-00-05 |   3   |    0     | voted(-)   | 00-00-00-01-00-05 |    0    |    0     |    0     | 2.000000
<0000000100006> /RRM/ <RRMT>: ** 00-00-00-01-00-05 | 00-00-00-01-00-06 |   3   |    0     |            | 00-00-00-01-00-05 |         |          |          | **
<0000000100006> /RRM/ <RRMT>: 00-00-00-01-00-06 | 00-00-00-01-00-05 |   3   |    0     | collecting | 00-00-00-01-00-06 |    0    |    0     |    0     | 3.999359
<0000000100006> /RRM/ <RRMT>:                                                                      pending... | 00-00-00-01-00-05
<0000000100006> /RRM/ <RRMT>:                                                                      pending... | 00-00-00-01-00-07
<0000000100006> /RRM/ <RRMT>: 00-00-00-01-00-07 |                   |   4   |    0     | collecting | 00-00-00-01-00-06 |    0    |    0     |    0     | 3.999359
<0000000100006> /RRM/ <RRMT>:                                                                      pending... | 00-00-00-01-00-06
<0000000100006> /RRM/ <RRMT>:                                                                      pending... | 00-00-00-01-00-07
<0000000100006> /RRM/ <RRMT>:  +------------------+-------------------+-------+----------+------------+-------------------+---------+----------+----------+---------
```

Listing 7.5: A Query is Received and Rejected

The answer to this problem is given by two additional tools, which have also been designed and implemented in the context of this work: `combine` and `verify`.

Combining Local Graphs

The former takes the configuration script and extracts all `_rrm_storegraph` directives, so as to retrieve the filenames of local RRM graphs. Then, it starts parsing the graph files, one after the other, adding each vertex and each edge it encounters to a new global, combined graph. While adding graph elements, the tool ascertains that the same vertex or edge has not been assigned different colors in different local graph files. If color assignments are not ambiguous, a combined graph is written to the hard disk, which can be visualized in `gview`, for example. Running the graph combiner yields an output like this:

```
Graph Combiner for The Wireless Wide Area Network Simulator Version 1.0
Copyright(C) 2002-2003 University of Wuppertal, Germany. All rights reserved.

Reading script "config10c-rrm.txt"... done. 10 graphs identified.

Processing rrm-graph-v000000010000.txt... ok. 4 vertices, 6 edges.
Processing rrm-graph-v000000010001.txt... ok. 5 vertices, 8 edges.
Processing rrm-graph-v000000010002.txt... ok. 6 vertices, 10 edges.
Processing rrm-graph-v000000010003.txt... ok. 7 vertices, 12 edges.
Processing rrm-graph-v000000010004.txt... ok. 7 vertices, 12 edges.
Processing rrm-graph-v000000010005.txt... ok. 7 vertices, 12 edges.
Processing rrm-graph-v000000010006.txt... ok. 7 vertices, 12 edges.
Processing rrm-graph-v000000010007.txt... ok. 6 vertices, 10 edges.
Processing rrm-graph-v000000010008.txt... ok. 5 vertices, 8 edges.
Processing rrm-graph-v000000010009.txt... ok. 4 vertices, 6 edges.
Combined graph consists of 10 vertices and 18 edges.
Storing graph "config10c-rrm-combined.txt"... success.
```

Verifying Graph Integrity

The combined graph may also be further processed using the `verify` tool. This tool reads any graph (may it be a local RRM graph or the combined graph), and looks for potential constraint set violations. Without such a tool, it would be almost impossible to assess whether TBSD-RRM has really produced a set of assignments, which respects the constraint set even in the global view. Notice that the debug version of WWANS always verifies integrity of local RRM tables, after any modification (table fusion). In addition, this tool is able to count the number of unique colors that have been assigned, and the highest color index that has been used. What is more, it is possible to let `verify` create color histograms, indicating how often each color has been assigned. For the previous example, the following output is obtained:

```
Verify Constraints for The Wireless Wide Area Network Simulator Version 1.1
Copyright(C) 2002-2004 University of Wuppertal, Germany. All rights reserved.

Reading graph "config10c-rrm-combined.txt"... done.

Graph consists of 10 vertices and 18 edges.
Average degrees: in = 1.8, out = 1.8, total = 3.6.
Maximum degrees: in = 2, out = 2, total = 4.
Maximum degree ratio = 1
18 edges out of 18 (100.00%) have got a color.
8/8 colors have been assigned.
Storing color histogram "histogram.txt"... done.

Checking graph integrity... done.
The graph is in accordance with the constraint set.
```

7.3.4   Comparing Assignment Results

Of course, it is desirable to have an implementation of the UxDMA algorithm in its original, centralized form. Such an implementation has been created and a tool called `unidca` provides the desired functionality. It accepts a textual graph description of the network topology as input. Then it runs UxDMA with the specified constraint set, target (either edge or vertex), and ordering heuristic. The result is stored as a graph file, exactly in the same format that local TBSD-RRM graphs use. Refer to listing 6.2 for an example. Consequently, `gview` and `verify` can be used to visualize assignments and gather statistics, respectively.

7.4   Third Example: A WSDP Flow

The script in listing 7.6 has been used to obtain the results, which have been presented earlier (chapter 5). Again, a new software tool dedicated to the preparation of sensor update files has been created, called `wsdpprep`. This tool accepts a reference value stream and eliminates sensor samples, in order to create a set of updates, which are actually passed to WSDP for transmission. As mentioned in the previous chapter, a server application will process the update file, in turn. The preparation tool supports those three methods of data reduction at the source, which have been presented in chapter 5: uniform, random and adaptive sub-sampling.

```
// config.txt : setup script
//
// Testbench for Wireless Sensor Data Protocol (WSDP)

topography
{
    station
    {
        position = (-550.0, 550.0, 1.2);

        protocols
        {
            attach(ieee_802_11_phy(2.4e9, 11e6, 2000000));
            attach(ieee_802_11_mac("00-00-00-ff-00-01"));
            attach(ieee_802_2_llc);
            attach(ipv4("192.168.1.1", "255.255.255.0", "0.0.0.0"));
            attach(udp);
            attach(wsdp(8500));
        }
    }

    station
    {
        position = (550.0, 550.0, 1.2);

        protocols
        {
            attach(ieee_802_11_phy(2.4e9, 11e6, 2000000));
            attach(ieee_802_11_mac("00-00-00-ff-00-02"));
            attach(ieee_802_2_llc);
            attach(ipv4("192.168.1.2", "255.255.255.0", "0.0.0.0"));
            attach(udp);
            attach(wsdp(8500));
        }
    }

    station
    {
        position = (0, 0, 1.4);

        protocols
        {
            attach(ieee_802_11_phy(2.4e9, 11e6, 2000000));
            attach(ieee_802_11_mac("00-00-00-ff-00-03"));
            attach(ieee_802_2_llc);
            attach(ipv4("192.168.1.3", "255.255.255.0", "0.0.0.0"));
            attach(udp);
            attach(wsdp(8500));
        }
    }
}

applications
{
    // Create a sensor data client application
    wsdp_client("192.168.1.1", 8.0, 24.0, 0.01,
        8086, 2, "sensor-results.txt");

    // Create a sensor data client application
    wsdp_client("192.168.1.2", 8.0, 24.0, 0.01,
        8086, 3, "sensor-results-2.txt");

    // Create a sensor data server application
    wsdp_server("192.168.1.3", "255.255.255.255",
        8086, 8.0, "sensor-updates.txt");
}

events
{
    // Turn on stations simulataneously
    power("00-00-00-ff-00-01", 1, 0.0);
    power("00-00-00-ff-00-02", 1, 0.0);
    power("00-00-00-ff-00-03", 1, 0.0);

    // Turn off stations simulataneously
    power("00-00-00-ff-00-01", 0, 30.0);
    power("00-00-00-ff-00-02", 0, 30.0);
    power("00-00-00-ff-00-03", 0, 30.0);
}
```

Listing 7.6: Setup Script used for Simulating WSDP

Part III

Demonstrator Design

This page intentionally left blank.

## CHAPTER 8

Transceiver Architectures

Starting with a brief review of classical and modern transceiver architectures, this chapter prepares the theoretical background for the prototype transceiver to be presented afterwards. For clarity's sake, it should be emphasized that an architecture is sought, which provides a flexible development environment. That is, criteria like ease of reconfigurability, shortness of development cycle times, a potential for high performance etc. take precedence over size, power consumption, and the like.

8.1 Classical Approaches

Classical receiver designs will be in the focus of the following sections. There exists a corresponding transmitter design for each receiver architecture, which consists of essentially the same fundamental building blocks.

8.1.1 Direct Conversion Architecture

The direct conversion receiver, also known as homodyne or zero-IF receiver, is the most natural approach to down-converting an RF signal directly to baseband. After the preselection filter, a low noise amplifier (LNA) drives a quadrature mixer, which down-converts the RF signal to baseband. The local oscillator (LO) frequency is set equal to the center frequency of the desired channel. Variable gain amplifiers (VGAs) in the baseband in-phase (I) and quadrature (Q) branches are used to drive the subsequent analog-to-digital converters (ADCs) just below full-scale. Automatic gain control (AGC) ensures optimal dynamic range performance. Finally, baseband processing

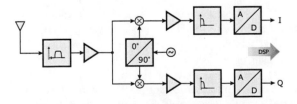

Figure 8.1: Homodyne Receiver

143

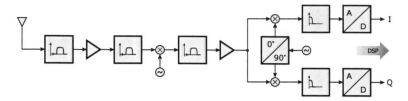

Figure 8.2: Heterodyne Receiver

is done in software [115] running on digital signal processors (DSPs), possibly with dedicated co-processors for turbo and VITERBI decoding, etc.

The homodyne architecture offers a number advantages: low complexity, amenability to monolithic high-integration, and lax filtering requirements. But it has also numerous disadvantages: the LO must provide two output signals, which are accurately in phase quadrature and amplitude balance over the entire input signal frequency range. Besides I/Q mismatch, LO-leakage causes time-varying DC-offsets [116]; $1/f$ noise is a major problem, too.

8.1.2 Conventional Heterodyne Architecture

Most of the transceivers used to date are based on a conventional heterodyne architecture, as shown in figure 8.2. The preselection filter immediately following the antenna, removes out-of-band signal energy and partially rejects image band signals. The LNA provides modest RF gain at a low noise figure, and is used to drive an image-reject filter, which further attenuates spectral components in the undesired image band. The entire band of interest is subsequently down-converted to an intermediate frequency (IF) by mixing with the output of an LO. Often, active mixers are used which also provide some additional gain. An IF channel select filter isolates the desired channel and reduces distortion and dynamic range requirements of the remaining receiver parts. The resulting IF signal is passed to a VGA, to ensure proper scaling in front of the ADCs. Again, baseband processing is the DSP's task.

Advantages of this architecture include its good selectivity and sensitivity. Moreover, gain is distributed over several amplifiers operating in different frequency bands, which reduces the potential for instability. Compared to the homodyne receiver's LO requirements, a phase quadrature and amplitude balanced LO is only required at a single frequency. Disadvantages include high complexity, possible requirement of several LO signals, and high-quality ceramic or SAW IF filters. This makes it nearly impossible to achieve single chip realizations.

8.2 Software Radio

The two previously presented classic architectures use digital signal processing only at baseband. The more recent idea of "software radio" [117] aims to push the ADC as close as possible toward the antenna. The ideal goal would be having a receiver consisting of no more than antenna, LNA and bandpass sampling ADC. All other functions (such as down-conversion, out-of-band interference rejection, etc.) would be implemented in the digital domain, preferably as software modules executed by a general purpose processor (GPP). What makes this approach so attractive is its inherent flexibility: a single radio could be used to for different air standards by running different software on the GPP.

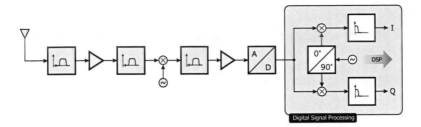

Figure 8.3: Digital IF Receiver

8.2.1 Software Defined Radio

Today, such an approach is technically not feasible for frequencies above a few megahertz, per-haps. Anyway, it is possible to make much of the radio software configurable, or software defined: design choices are not restricted to "all or nothing". In a software defined radio (SDR) [118], as much as possible is done using a GPP, preferably programmed in a high-level programming language, such as C++. More time critical portions can be implemented using native assembler code on the same GPP: this improves execution speed at the cost of much reduced portability, increased development effort, etc. When GPP performance becomes a bottleneck, dedicated circuitry becomes a necessity. But even such dedicated circuits, tailored to just a single specific application, may be configurable: as an example, consider an FIR filter realized as an application specific integrated circuit (ASIC), where filter coefficients are stored in programmable registers.

Among other benefits, SDR technology enables multi-mode applications: a single handset may be reconfigured dynamically to join a cellular network, a wireless LAN, or to spontaneously form an ad hoc network with co-located stations. In such a manner, diversified communication possibilities are opened. Imagine the following: when at home, a user may form an ad hoc network with some of his friends in the neighborhood, and play an interactive online game. Another time, the user may be driving in his car without a suitable ad hoc network in reach. Then, there is still the possibility of joining a GSM network to make a phone call.

8.2.2 State of the Art: Digital Intermediate Frequency

Recent advances in semiconductor technology have made so called "digital IF" [119] systems a reality. ADCs with sampling rates in excess of 100 MSps, 14/16 bits of resolution and input bandwidths up to 270 MHz allow bandpass sampling of IF signals up to 200 MHz. Digital-to-analog converters (DACs) with input rates up to 160 MSps and $2\times/4\times/8\times$ interpolating FIR filters provide direct IF reconstruction capabilities[1]. These converters enable cost-effective implementations of SDRs, based on digital signal processing (DSP) techniques. As shown in figure 8.3, this design resembles the heterodyne architecture – with the difference that I/Q separation and channel selection are done in the digital domain. Naturally, this approach also creates severe challenges for the design of a suitable RF front end, depending on "how much" of the RF spectrum shall be processable. Anyway, the present work's scope reaches no further than just beyond DAC/ADC, where the transition to a suitable analog front-end (AFE), like the RF stage outlined in [120], takes place.

---

[1]Notice that these are the technical data of the ADC and DAC components actually used in the prototype demonstrator, which is going to be outlined in the next chapter. It can be expected that performance in terms of resolution, input bandwidth, sampling rates, etc. steadily increases while cost decreases

Digital Front End

As already mentioned before, baseband processing (coding, modulation, etc.), has already been done in the digital domain for quite a time. However, it was (and still is) not so easy for a GPP or general purpose DSP to achieve the data rates required to create the first IF digitally. For instance, if a 40 MHz pass-band was desired, the DSP would have to drive the DAC with at least 80 MSps and 12 to 16 bits of resolution. Analogously, in the receiver case, the DSP would have to accept 80 MSps from the ADC.

This necessitates utilization of ASICs for the implementation of the digital front end (DFE) [121]. Nowadays, up-conversion from baseband to a first IF has also become feasible. The same is true for the dual operation, down-conversion from IF to baseband. The necessary rate change [122] is usually performed in a staggered fashion using finite impulse response (FIR) and cascaded integrator comb (CIC) filters. These are employed as interpolating filters for up-conversion, while the same structures are used as decimating filters for down-conversion. Staggered filtering accounts for the fact that highly selective, high-order, high-complexity filters are difficult to implement at sufficiently fast clock speeds. In the down-conversion case, for example, the sampling rate is successively reduced in each stage. This way, low-complexity decimating filters can be used at high clock speeds, while more selective filters can be used at lower data rates. In addition to rate conversion, a tuning capability is also required. Digital tuners comprise a numerically controlled oscillator (NCO) and a quadrature amplitude mixer (QAM). Up- and down-conversion circuits are usually implemented as ASICs or field programmable gate array (FPGA) [123] designs. The demonstrator prototype, which is going to be detailed in the next section, implements the DFE using custom off-the-shelf components. Semiconductor companies like Texas Instruments (previously Graychip), Intersil, Analog Devices, and others offer suitable devices under different names, e.g. digital up-/down-converter, transmit/receive signal processor (TSP/RSP), etc.

## CHAPTER 9

Transceiver Prototype

A new hardware platform for research and development is presented, which is capable of multi-carrier and multi-mode operation. It provides a fully digital solution up to the first IF at around 80 MHz. The platform has been designed to integrate with the simulator presented in part II, so as to provide a consistent environment for R&D in the emerging sector of multihop, wide area networks. Future wireless ad hoc networks with inherent QoS features, i.e. 4G+ systems, are in the focus of interest. Thanks to its software configurable nature, the transceiver is not only suitable for 4G, but it is also capable of carrier-processing for 2G/3G networks, such as GSM/EDGE and UMTS.

The presented solution is quite flexible, powerful and yet cost-effective, when compared to commercial system solutions. Furthermore, the modular architecture is extensible to support the latest technologies, such as: multiple-input, multiple-output (MIMO); space-time coding; beam-forming; predistortion etc. A wide-spread general purpose DSP assisted by FPGAs and dedicated transmit and receive signal processors works at the transceiver's core. In essence, it presents an implementation of the digital IF transceiver reviewed in the previous chapter. Research on wide area multihop ad hoc networks is still on-going. Thus, the new hardware platform must provide enough flexibility to support different air standards in terms of modulation and multiple access schemes, channel bandwidths, and so forth.

9.1   Introduction to the Architectural Concept

All in all, incorporation of a variety of technologies seems to provide the best mix between flexibility, cost, development cycle turn-around times, and performance, when designing a digital IF transceiver. For instance, much of the baseband signal processing is optimally done by a DSP, because of the unparalleled flexibility, portability and short development cycle times inherent to C++. The next grade of trade-off between flexibility and performance is achieved by assembler programming of time critical functions within the DSP. If the performance required by some computationally intensive algorithms is still not achievable, assistance by hardware circuits is required, for instance a dedicated co-processor for chip-rate processing of CDMA signals, fast VITERBI decoding, etc. Again, different grades of flexibility vs. performance vs. cost are available based on the chosen technology.

Using a hardware description language, such circuits may be realized as FPGA designs. This requires a whole lot of extra effort and expertise compared to programming in C++, though virtually any digital circuit may be implemented, easily modified and upgraded later on. However, certain functions may still be difficult to implement with the required clock speeds due to insufficient FPGA resources (logic blocks, wiring). Then, the flexibility vs. performance competition is won by programmable ASICs. They have a fixed circuit structure, but are still configurable using programmable registers. Two such devices are used in the presented transceiver design for digital up- and down-conversion. They implement the TX and RX filter masks as specified in various standards (like GSM or UMTS), perform the necessary rate change between baseband and IF, and tune to the desired channel (frequency translation). Finally, digital-to-analog and analog-to-digital conversion is done by ASICs without any programmability at all, and further analog components are required for IF/RF conversion.

9.2   General Purpose Digital Signal Processor Board

The novel hardware demonstrator prototype employs a general purpose DSP for baseband processing, control and monitoring of the other system components. Three DSPs from Texas Instruments can be used interchangeably: two floating point DSPs, and a fixed point DSP. All comprise a very long instruction word (VLIW) C6000 processor core. The VLIW processor architecture allows many instructions to be issued in a single clock while still allowing for very high clock rates. The DSPs mentioned below are able to execute up to eight instructions in parallel. This architecture can achieve extremely high processing rates but puts more burden on the compiler to schedule concurrently executed instructions. Highly optimizing code generation tools are provided to relieve as much of the optimization burden from the programmer as possible.

9.2.1   Development Starter Kit

To shorten development time, a Texas Instruments Development Starter Kit (DSK) has been used as a starting point. The DSK hosts a DSP together with some memory and peripherals for audio processing. The DSP's most important signals (memory bus, serial ports, interrupt lines, etc.) are made available to users via two expansion interfaces: one for additional memory or memory mapped devices, the other for additional peripherals. Originally, this expansion interface had been introduced with the TMS320C6701 evaluation module. Today, a number of different DSK models are available, and each model offers a compatible expansion interface. During this work, three different DSK models have been used as part of the novel SDR implementation interchangeably:

- Development started with a DSK model hosting a TMS320C6711 floating point DSP running at 150 MHz. In addition, 16 MB of 100 MHz SDRAM, 128 KB of flash ROM, and a 16-bit, 8 kSps audio interface are available. The board is connected to a PC's parallel port for development and debugging purposes. Although this DSK may still be used, it is deemed outdated.

- A new DSK model has recently become available, hosting the latest member of Texas Instruments' floating point DSP family, a TMS320C6713 running at a core frequency of 225 MHz. The memory configuration comprises 192 KB of internal DSP memory, 8 MB of SDRAM and 512 KB of flash ROM. Furthermore, the audio interface has been improved to support 16–32 bits of resolution at sampling rates of 8–96 kSps. This DSK model is connected via Universal Serial Bus (USB).

Figure 9.1: Prototyping System, Third Revision – Photograph of tntSDR Mounted on DSK

- The third DSK also belongs to the class of more recent USB boards with improved audio interface. This DSK model offers the same equipment as the previous one, except for the DSP, which is a 600 MHz TMS320C6416 with 1 MB of internal memory. Moreover, 16 MB of SDRAM are available.

9.3 A Novel Daughter-Board: tntSDR

During this work, a novel daughter-board has been designed, which may be mounted on top of the DSK. It connects to both expansion interfaces. The new daughter-board is entitled "tntSDR". Both boards together, i.e. DSK & tntSDR, form a very flexible and powerful digital IF transceiver testbed. Naturally, hardware development is often tedious, and usually a number of revisions are required to obtain the desired results. It is the third revision of tntSDR [124], which is presented here. A description of previous revisions is provided in [78, 125].

Figure 9.1 shows a photograph of the tntSDR daughter-board mounted on top of the TMS320C6416 DSK. An architectural overview is provided in figure 9.2. Here, DSK components are identified by gray boxes with a thick frame on the left-hand side of the illustration; the remaining components belong to the daughter-board. The third revision daughter-board holds the following major components:

- AD6623 four-channel 104 MSps transmit signal processor (TSP)

- AD6624 four-channel 80 MSps receive signal processor (RSP)

- AD6645 14 bit 80/105 MSps bandpass-sampling ADC

- AD9772A 14 bit 160 MSps 2× interpolating "direct IF" DAC

- XC2S200 Xilinx Spartan II FPGA (2×)

- XC18V04 in-system-programmable 4 Mb flash ROM

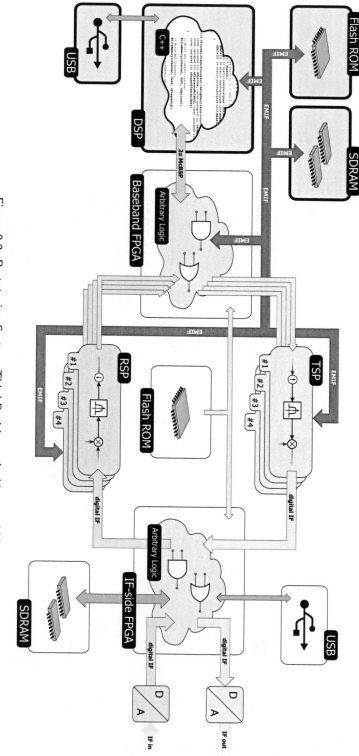

Figure 9.2: Prototyping System, Third Revision – Architectural View

- FT245BM USB FIFO

- MT48LC8M8A2 SDRAM ($4 \times 8 \times 8$ Mb $= 32$ MB)

It also comprises a MAX3323E RS-232 line driver/buffer for backwards compatibility.

9.3.1 Role of the General Purpose DSP

The DSK board is connected to a host PC, which runs the Code Composer Studio software. This feature-rich software suite provides an integrated development environment for Texas Instruments DSPs. Software development starts with a description of the hardware configuration. This way, the compiler learns for which particular DSP architecture the C++ or assembler source code is intended; and the linker learns address space constraints, memory mappings (internal memory size, SDRAM address space, ...), etc. The compiled and linked binary output file can be immediately loaded into a DSK board via parallel port or USB. In addition, JTAG programming is also an option for proprietary hardware designs. But it is also possible to create a bootable image, which can be stored in the DSK's flash ROM. This allows a stand-alone operation of the DSK.

A lot of the flexibility, ease of use and shortness of development cycle times is due to this combination of hard- and software. Its only a matter of seconds to modify the source code, rebuild the project and download the binary image into the DSK. In addition, it is possible to access the entire DSP memory space from inside Code Composer Studio. This way, certain memory regions can be inspected, and specific values may be modified. Moreover, data can be exchanged between host PC and DSK at runtime. This is a very useful debugging path. Besides plain text output from the DSP, which will be displayed in real-time in Code Composer Studio, it is also possible to transfer arbitrary binary files. Later on, two novel diagnose tools, ad6623chk and ad6624chk, will be introduced, which make use of this data path.

During this work, a suite of C++ classes has been developed for the DSK. This piece of software runs on the three boards mentioned earlier and allows controlling TSP and RSP components. In addition it provides an interface to the baseband FPGA. Currently, the DSP controls transceiver initialization at startup, programs filter coefficients, desired NCO frequencies, rate change factors etc. into RSP and TSP components. Moreover, it provides baseband data, either as I/Q symbol vectors, or as raw binary streams. This depends on the actual baseband FPGA design. In the present configuration, the DSP's 32-bit wide external memory interface (EMIF) is operated at 100 MHz. It supports different memory architectures, such as SDRAM, ASRAM, SBSRAM, and flash ROM simultaneously. These can be operated in a time-multiplex manner. For example, the SDRAM and flash ROM memories on the DSK are connected to the EMIF.

9.3.2 Baseband FPGA

A Spartan II FPGA is also directly connected to the DSK's 100 MHz 32-bit wide EMIF. A VHDL design has been created, which allows mapping portions of the FPGA's internal block RAM (BRAM)[1] into the DSP's memory space as ASRAM. This memory-mapped device approach features an extremely high-throughput path, which is able to serve even very challenging requirements. In addition, two of the DSP's high-speed bidirectional synchronous serial ports are also connected to the baseband FPGA. These multichannel buffered serial ports (McBSPs) provide separate clock, framing and data signals. These McBSPs are capable of clock speeds in excess of 100 MHz, depending on the DSP type and its internal core frequency[2].

---

[1]BRAMs are very fast synchronous dual-port static RAM blocks inside Xilinx FPGAs
[2]TMS320C6711 DSK: 75 MHz; TMS320C6713 DSK: 112.5 MHz; TMS320C6416 DSK: 150 MHz

Arbitrary digital designs can be implemented in the baseband FPGA. Besides a number of quite simple functions, the baseband FPGA may also be used to implement more demanding approaches. Simple functions include routing of McBSP signals to TSP and RSP baseband ports, and providing the necessary "glue" logic between DSP and TSP/RSP processors. This glue logic includes clock management, address decoding, reset signal generation, etc. Data reformatting is another simple function that can be implemented in VHDL.

More complex functions could include convolutional coding and VITERBI decoding, for example. These would be ideal candidates for a high-performance digital implementation. Obviously, implementing these parts as dedicated circuits takes a lot of the computational burden from the DSP, saving resources for other tasks. The promising approach of cut-through switching can be verified by realizing a cross-bar switch in the baseband FPGA. This would be of particular interest for high-performance, QoS-aware multihop ad hoc networks.

A further very natural function to be implemented inside the baseband FPGA is chip rate processing for CDMA systems. With the DSP alone, it would be impossible to achieve the required performance. Just consider path detection in a rake receiver: in each rake finger a 256-tap FIR filter matched to the desired synchronization code is required. In addition to path detection, a rake receiver comprises several other functional components: channel estimation and derotation (once per finger), maximum ratio combining, despreading, etc.

A WCDMA transmission chain including transmitter and rake receiver has been designed for tntSDR [126]. Basic modules (correlator, spreader, despreader, etc.) can be easily reused for air standards other than WCDMA. The transmitter consists of spreader, GOLD code generator, scrambler, pilot sequence generator, and power control units. The rake receiver is configurable with respect to the number of fingers, oversampling rate, etc. It comprises optimized, hand-crafted VHDL modules for input signal normalization, despreading, path detection, channel estimation, derotation, slot and frame synchronization, etc. It is quite obvious that potential baseband designs are too numerous to mention them all.

9.3.3 Transmit Signal Processor

The TSP's principle task is to create high bandwidth data for the DAC from baseband data provided by the DSP or the baseband FPGA. It is suitable for synthesis of multi-carrier, multi-mode digital signals. Typical applications include cellular base stations for GSM, EDGE, WCDMA, IS-95/136, cdma2000, beam forming antenna arrays, and others. The AD6623 contains four identical TSPs along with synchronization circuitry in one package. Besides multi-carrier operation, this allows combining of two or more channels for increased processing power. In other words, channels can be combined to obtain more selective filters. A single AD6623 chip is able to process four GSM/EDGE carriers or two UMTS carriers (two TSP channels per carrier) simultaneously. Furthermore, several chips may be cascaded to process 8, 12 or 16 carriers with a single DAC. Vice versa, a single AD6623 can drive two DACs independently from each other. Each of the four parallel TSP processing chains consists of a number of programmable cascaded elements. The most important ones are depicted in figure 9.3:

- **Serial Input Port.** The high-speed synchronous serial port accepts baseband data from the baseband FPGA. Two data types may be used for this purpose: either I/Q vectors or symbol bits. Symbol bits are used when the TSP's modulator shall be used to map symbol bits to I/Q constellations. The TSP is able to modulate $\frac{\pi}{4}$-DQPSK, $\frac{3\pi}{8}$-8-PSK, 8-PSK, QPSK, and GMSK directly from symbol bits. If other modulation schemes are required, e.g. 16-QAM, I/Q vectors must be passed to the TSP. In this case, the DSP or a suitable design in the baseband FPGA must perform constellation mapping.

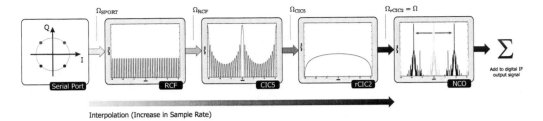

Figure 9.3: Signal Path of a Single TSP Channel (Four Available)

- **Interpolating RAM coefficient filter (RCF).** The RCF implements any of the modulations mentioned above or an interpolating FIR filter with programmable coefficients. A polyphase implementation avoids unnecessary calculations for zero-valued input samples. Such samples are used to increase the data rate (zero stuffing). It also implements an all-pass phase equalizer which meets the requirements of IS-95, if desired. In addition to pulse shaping, the RCF filter can be used to compensate for the passband droop of subsequent elements in the overall transmission chain: from CIC5 and rCIC2, over DAC, analog filters and mixers to the high power amplifier.

- **Scale and power ramp.** This block allows power ramping on a time-slot basis, as it is required by GSM/EDGE, for example. A fine scaling unit allows an easy amplitude adjustment on time-slot basis, too. With this feature, it is possible to reduce the amplitude of undesired spurious signals. These occur, when the DAC converts a rapid series of on/off data bursts to produce an analog output signal. This feature can be used to ramp down, pause for a certain guard interval, and ramp up at the beginning of the next time slot. Modulation is adaptable on a burst-by-burst basis, i.e. different modulations can be used in subsequent time slots (e.g. GMSK then 8-PSK, then again GMSK).

- **Fifth order cascaded integrator comb interpolating filter (CIC5).** The CIC5 provides integer rate interpolation from 1 to 32 and coarse anti-image filtering. For all interpolation rates, the CIC5 has a finite impulse response with perfect phase linearity.

- **Second order re-sampling CIC filter (rCIC2).** The rCIC2 provides fractional rate interpolation from 1 to 4096 in steps of $1/512$. The resampling capability allows using a single master clock, which is currently 80 MHz for the tntSDR, to achieve a wide range of baseband symbol rates with non-integer digital IF to symbol rate ratios. Notice that the baseband symbol rate is determined by dividing the digital IF sampling rate of 80 MSps by the overall channel interpolation factor.

- **Numerically controlled oscillator/tuner (NCO).** The tuner consists of a 32-bit quadrature NCO and a quadrature amplitude mixer (QAM). Here, the NCO serves as an LO and the QAM tunes the interpolated complex baseband signal from rCIC2 to the NCO frequency. Advanced phase and amplitude dither options provide a means to improve the NCO's spurious performance. In general, the worst case spurious signal from the NCO is better than -100 dBc for all output frequencies. In addition to the NCO frequency, a phase offset may be specified as well. This way, multiple NCOs can be synchronized to synthesize sine waves with a known phase relationship.

Overall Transfer Function

In each filter stage, the data rate is increased by zero-stuffing. For instance, when the RCF interpolates by a factor of L_{RCF} inserting $L_{RCF} - 1$ zero-valued samples prior to FIR filtering, its output data rate Ω_{RCF} equals L_{RCF} times its input data rate Ω_{SPORT}. In the TSP's data sheet, transfer functions for each filter stage are given with respect to their individual output frequencies. For the RCF block, the transfer function is given by $\mathcal{Z}$-Transform of $h(n), n \in [0, N_{RCF} - 1]$, where $h(n)$ are the programmable impulse response coefficients:

$$H_{RCF}(z) = \sum_{n=0}^{N_{RCF}-1} h(n) z^{-n} \tag{9.1}$$

The RCF is implemented as an efficient polyphase filter, in order to avoid unnecessary multiply and accumulate (MAC) operations, when input samples are zero-valued. Remember that $L_{RCF} - 1$ input samples are definitely zero-valued because of zero-stuffing. Consequently, when programming the chip, coefficients must be stored in polyphase order.

The CIC5 filter is characterized by

$$H_{CIC5}(z) = \left(\frac{1 - z^{-L_{CIC5}}}{1 - z^{-1}} \right)^5, \tag{9.2}$$

with L_{CIC5} being the programmable interpolation factor for this filter stage. Because of its simplicity compared to the highly configurable RCF, the CIC5 can run at much higher clock speeds – which is the motivation for the staggered filter design. Finally, the even simpler rCIC2 filter runs at the TSP's full clock speed, i.e. at up to 104 MHz. In contrast to the CIC5, it does not only provide interpolation by a factor of L_{rCIC2}, but also decimation by M_{rCIC2}. Its transfer function is given by

$$H_{CIC2}(z) = \left(\frac{1 - z^{-\frac{L_{rCIC2}}{M_{rCIC2}}}}{1 - z^{-1}} \right)^2. \tag{9.3}$$

Now, combining these individual transfer functions into a single overall transfer function $H\left(e^{j\omega T}\right), T = \frac{2\pi}{\Omega}$, where $\Omega = \Omega_{rCIC2}$ identifies the TSP's output sample rate, we obtain:

$$
\begin{aligned}
H\left(e^{j\omega T}\right) = {} & \frac{1}{L_{RCF}} \sum_{n=0}^{N_{RCF}-1} h(n)\, e^{-jn L_{CIC5} \frac{L_{rCIC2}}{M_{rCIC2}} \omega T} \\
& \cdot \frac{1}{L_{CIC5}} \left(\frac{1 - e^{-j L_{CIC5} \frac{L_{rCIC2}}{M_{rCIC2}} \omega T}}{1 - e^{-j \frac{L_{rCIC2}}{M_{rCIC2}} \omega T}} \right)^5 \\
& \cdot \frac{M_{rCIC2}}{L_{rCIC2}} \left(\frac{1 - e^{-j \frac{L_{rCIC2}}{M_{rCIC2}} \omega T}}{1 - e^{-j\omega T}} \right)^2
\end{aligned}
\tag{9.4}
$$

Two actually implementable example filters are shown in figure 9.4. One is representative for 3G wideband communications (figures 9.4(a)/9.4(b)), while the other demonstrates 2G narrow-band suitability (figures 9.4(c)/9.4(d)). The sampling period T in both illustrations corresponds to the system's master clock frequency of 80 MHz. In other words, normalized frequencies $-\pi \le \omega \le \pi$ map to the frequency range -40 MHz $\le f \le 40$ MHz.

Coefficients shown in figures 9.4(b)/9.4(d) and parameters detailed in table 9.1 correspond to reference designs outlined in Analog Devices application notes. Figures 9.4(a)/9.4(c) show

Parameter	WCDMA Filter	EDGE Filter
L_{RCF}	6	6
N_{RCF}	24	96
L_{CIC5}	8	20
L_{rCIC2}	1	2
M_{rCIC2}	1	1

Table 9.1: Parameters used for Reference Filters

how the different stages work together in order to create a suitable filter function that matches specification requirements. In particular, this view shows how the CIC filters are used to null-out undesired images created by interpolation. For clarity's sake, a view on individual contributions to the overall transfer functions, as well as the resulting transfer functions, are shown in figure 9.5.

The outputs of the four individual TSP channels are summed and scaled on-chip in a final stage. The resulting wide-band signal is passed to the IF-side FPGA, which delivers it to the DAC for reconstruction. The chip is managed and controlled through its microport interface, which connects to the DSP's EMIF. As a result, the TSP appears to the DSP as a memory mapped device. Baseband data comes either from one of the two DSP's McBSPs, or is provided by the baseband FPGA. When symbol rates are not very high (e.g. EDGE with 270 kSps), baseband data may be provided by the DSP. Conversely, a VHDL design inside the baseband FPGA would provide WCDMA I/Q vectors with a rate of 3.84 kSps.

A Tool for Verification of Correct TSP Programming

The TSP offers a wealth of registers, which are used to store filter coefficients, interpolation rates, NCO frequencies, phase offsets, etc. A single ill-programmed bit can cause severe malfunction of the entire channel. Moreover, the TSP is extremely amenable to overflows due to the underlying fixed point arithmetics. Filter coefficients and scaling factors must be well-designed, so as to avoid such unwanted behavior.

Therefore, a novel software tool has been developed which verifies correct programming of the TSP. This diagnose tool is called `ad6623chk`. It accepts a raw memory image of the TSP's registers and creates a human readable report, stating the various global and per-channel settings for all four TSP channels. In addition, the tool is able to export filter coefficients in ASCII format, which can be read by third party software. For instance, MATLAB can be used to create frequency/phase plots of the filter transfer function, or stem plots of the impulse response. This new tool also checks for constraint violations (allowable interpolation rates, number of RCF coefficients, etc.) and alerts about potential overflows. A fragment of such a report is printed in listing A.1. This fragment gives an impression of the numerous registers available for configuration.

9.3.4 Receive Signal Processor

The RSP has essentially the same structure as the TSP, with the difference that the signal processing chain is in reverse order compared to the TSP. Instead of interpolating filters, decimating filters are being used. The RSP accepts parallel wide-band input data from the IF-side FPGA. Usually this data comes from an ADC, but it is also possible to use perfect digital IF signals generated by the TSP, instead. The RSP selects a certain frequency band with its NCO/tuner. The NCO's output is processed by a second order resampling decimating CIC filter, which reduces the data rate in a first step. Following the rCIC2, a decimating CIC5 filter further reduces the

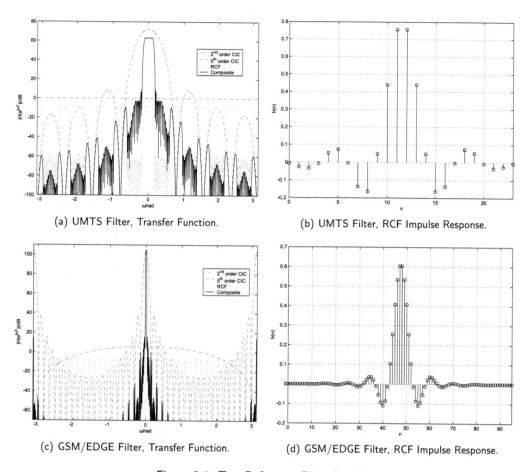

(a) UMTS Filter, Transfer Function.

(b) UMTS Filter, RCF Impulse Response.

(c) GSM/EDGE Filter, Transfer Function.

(d) GSM/EDGE Filter, RCF Impulse Response.

Figure 9.4: Two Reference Filter Designs

sampling rate, so as to allow the final decimating RCF to process more taps per output. The RCF's output is made available to the baseband FPGA via a dedicated serial port. Equal to the AD6623, which offers four TSPs in a single package, the AD6624 offers four individually configurable RSPs.

Another diagnose tool, very similar to ad6623chk, has been developed for the RSP. This tool, ad6624chk, operates in the same way as the TSP variant. It is able to translate a raw memory image into a human readable textual report, like the one presented in listing A.2. As for the TSP, filter coefficients can be converted from fixed point binary to floating point ASCII format. Distinctive to the RSP diagnose tool, raw baseband samples received by the RSP can also be converted to floating point ASCII format. All relevant RSP settings, such as the serial port word length and the binary output format, are taken into account for proper conversion.

A simple MATLAB script has also been developed, which can be used to visualize received I/Q symbols in a number of ways. For example, the script allows animating received samples in the complex plane, so as to trace constellations changes. In addition, another set of diagrams, which are also accessible in a static printed form, are also displayed by the script. In figure 9.6 three plots of the same recorded sequence are depicted. Here, the TSP has been used to create

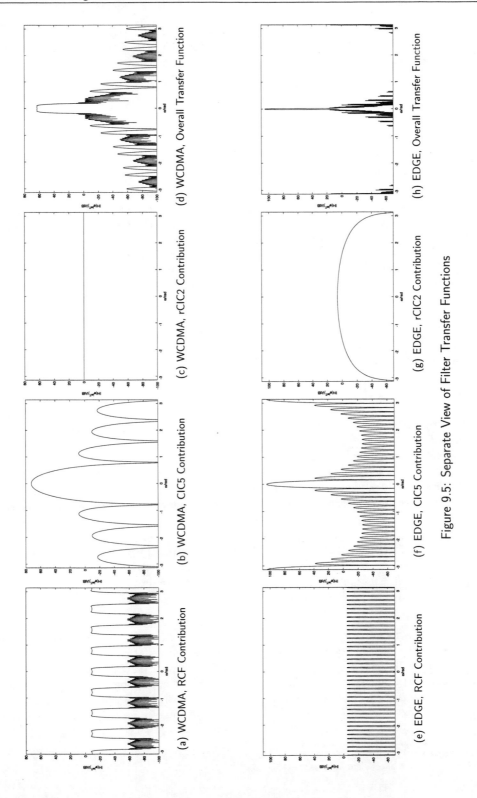

Figure 9.5: Separate View of Filter Transfer Functions

a four-constellation test data stream. Both, TSP and RSP, have been configured to employ the EDGE compliant filters mentioned earlier. Both have been tuned to the same IF channel. This test also proves correct operation of all components (i.e. DSP, baseband FPGA, TSP, IF-side FPGA, DAC, ADC, RSP), their interconnections, VHDL digital hardware designs and C++ software modules.

9.3.5   IF-side FPGA

A second Spartan II FPGA is inserted into the digital IF path. It is connected to the TSP's 18 bit digital output bus, as well as to the RSP's digital input bus. Hence, it is possible to route the TSP output to the RSP input, so as to allow receiver algorithms to be tested using high quality digital data. In addition, DAC and ADC are also connected to the IF-side FPGA. Using a software on the host PC, the RSP can process either TSP output, or ADC output. This switching can be done "on-the-fly". The IF-side FPGA is also connected to an USB FIFO, which provides a high-speed communication path with the host PC. This path can be used for data acquisition, monitoring and control. The digital IF signal synthesized by the TSP could be recorded into portions of SDRAM, which is also attached to the IF-side FPGA. Moreover, it would be possible to provide a reference IF signal to be decoded by the RSP, etc. It is also possible to record ADC output at the maximum sampling rate of 80 MSps. Such measurements could be used to evaluate multitone spurious free dynamic range (SFDR), signal-to-noise ratio (SNR), etc. An RS-232 serial port line driver is also available at the FPGA. It had been used in earlier revisions to deliver recorded data to the host PC.

High-speed Digital Data Acquisition

During early development phases, it is a good idea to capture samples of the digital IF signal synthesized by the TSP and evaluate them on a PC. The TSP provides a huge amount of programmable parameters, coefficients, and settings for each of its four channels – all of these must be handled with care. Failure to select and program appropriate values will result in a dramatic loss in spurious and noise performance.

   In order to evaluate the digital signal synthesized by the TSP without spectral components added by the DAC, a digital design for high-speed synchronous acquisition of the digital IF signal has been implemented [127] in VHDL. Upon request, it acquires the 16 most significant bits provided at the TSP's output at every 80 MHz clock cycle, and starts filling the Spartan II's on-chip BRAMs. The XC2S200 hosts 14 synchronous dual-port BRAMs of 4096 bits each, yielding a sample buffer of 3584 samples depth and 16 bits width[3]. Afterwards, the data is transmitted to a host PC running a newly developed acquisition and control software.

Acquisition Software   The mating software running on a PC, called sdr2scope, has been developed for the Windows operating system using C++ and the Microsoft Foundation Classes library. Figure 9.7(a) shows a screen shot. The software accepts data, either from the PC's serial or USB ports, and converts the raw binary data from 16 bit offset-binary or two's complement format into a floating point representation. This data is graphically visualized as a waveform – much like the signal an oscilloscope would display. Furthermore, a Fast FOURIER Transform is performed "on-the-fly", and results are also visualized – comparable to the image a spectrum

---

[3]In the future, the SDRAM can be used for increased depth and width of the memory buffer. A suitable SDRAM controller is already available as a VHDL design

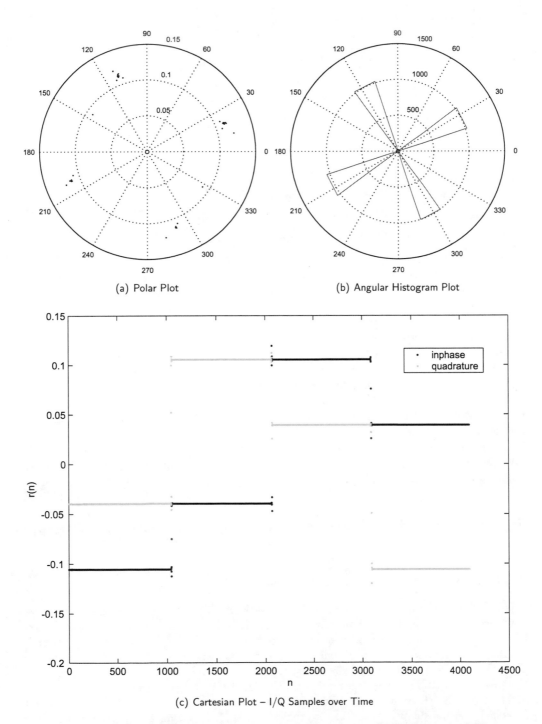

(a) Polar Plot

(b) Angular Histogram Plot

(c) Cartesian Plot – I/Q Samples over Time

Figure 9.6: Visual Representation of an I/Q Symbol Stream Received by the RSP

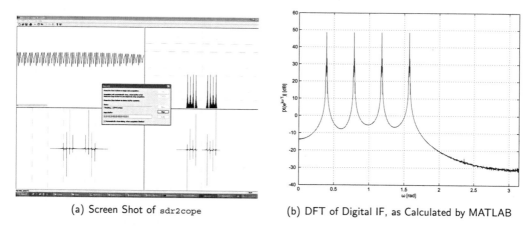

(a) Screen Shot of `sdr2cope`   (b) DFT of Digital IF, as Calculated by MATLAB

Figure 9.7: The Novel Acquisition and Control Software for tntSDR

analyzer would display. In addition, raw binary values and their resulting floating point representation may also be viewed in a list format for in-depth analysis. The floating point samples can be exported in ASCII format, too. This opens a path to almost every analysis software, e.g. Excel, MATLAB or custom solutions. The software implements common file operations as well, allowing raw data to be archived in its native format for later evaluation.

In figure 9.7(b), MATLAB has been used to perform a discrete FOURIER transform of the digitally captured IF signal. In this example, four streams of random baseband data have been filtered according to EDGE specifications, digitally up-converted, and used to modulate four carriers. Each carrier is spaced 5 MHz from its neighbor (assuming an 80 MHz master clock). Recalling stochastic signal processing theory, a system's transfer characteristic can be measured at the output by applying random noise to the input.

From this figure, one great benefit of a digitally synthesized IF signal becomes evident. Multiple carriers are simply added in the digital domain and jointly converted to analog. Only a single DAC is required, a single SAW filter, only one analog mixer, etc. Whereas the classical approach, where digital processing ends at baseband, would require two DACs (one for in-phase, one for quadrature) **per carrier** – plus additional SAW filters, mixers and all the analog circuitry required to build a multi-carrier transceiver. However, this poses several challenges to the design of suitable broadband RF components.

9.3.6   Analog to Digital Converter

Offering a broad input bandwidth, the AD6645 14-bit, 80 MSps wideband ADC is suitable for bandpass sampling of IF signals up to 200 MHz. It is a complete, high-performance, high-resolution conversion solution with on-chip voltage reference. For example, the converter is suitable for band-pass sampling of four WCDMA carriers in parallel. The AD6645 provides an SNR of 75 dB for a single 15 MHz input test tone, and 72 dB when the same tone is applied at 200 MHz. For a 70 MHz tone, it achieves an SFDR of 89 dBc. In the current revision of tntSDR, a standard SMA RF connector provides a single-ended analog input compatible with most test equipment. A transformer is used to turn the single-ended signal into a differential signal, as it is required by the AD6645. In the future, an IF band VGA with differential outputs could be connected immediately in front of the ADC. Its parallel output is passed to the baseband FPGA,

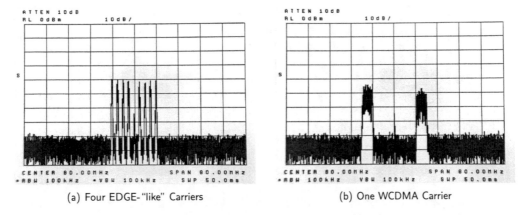

(a) Four EDGE-"like" Carriers   (b) One WCDMA Carrier

Figure 9.8: Measurements at the DAC Output

where it can be routed to the RSP, or recorded using the high-speed data acquisition.

9.3.7 Digital to Analog Converter

The AD9772A is a 14-bit 160 MSps 2× interpolating ADC capable of direct IF synthesis. The TSP's parallel output is routed through the IF-side FPGA to serve as an input to the ADC. The 2× digital interpolation filter can be configured for a high-pass response, thereby allowing direct IF conversion. The DAC also offers a zero-stuffing option to increase the DAC update rate by another factor of two, which enhances reconstruction performance. The tntSDR daughter-board provides jumpers, which can be used to configure the DAC mode. Usually, the DAC is operated in direct IF mode at an input sample rate of 80 MSps, with zero-stuffing enabled. This leads to a first IF around 80 MHz. The analog differential current output signal is converted to a single-ended signal for easy connection to test equipment. In the future, SAW filters, mixers, and frequency synthesizers will be required to produce a transmittable RF signal. The AD9772A provides a 67.5 MHz reconstruction passband when converting a 160 MSps input data stream. The related data sheet states a 74 dBc SFDR at 25 MHz. The 2× interpolating half-band FIR filter provides 73 dB image rejection with a 0.005 dB passband ripple.

Analog Measurements

To verify correct operation of the transceiver platform, a number of measurements have been conducted on the reconstructed analog signal. Notice, that these measurements are not meant to allow evaluation of the transceiver performance. In particular, the DAC has not been driven near to full-scale. Therefore, performance is clearly better than could be expected from the presented measurement results. A measurement performed using a spectrum analyzer is shown in figure 9.8. Here, random data has been provided to the TSP's input ports. Notice that the measurement is centered at 80 MHz. This demonstrates the DAC's direct IF capability.

9.4   Conclusion

In this chapter, a digital IF transceiver R&D platform has been presented. It is based on a hybrid approach consisting of a general purpose DSP, FPGAs and ASICs. This mix allowed to develop a functional transceiver in a relatively short period of time. This transceiver has been designed as a platform for research and development. Hence, flexibility and the capability to host a lot of different VHDL and C++ designs for verification and optimization purposes were primary criteria. The transceiver is able to support different air standards, which is an important metric with respect to emerging 4G systems.

In future work, this modular platform can be extended to actually implement a variety of new technological approaches. For example, a further DAC can be easily attached to tntSDR's user connector, which exposes more than 60 FPGA IOs. This way, it is possible to exploit transmit diversity, e.g. using ALAMOUTI's famous scheme [128]. Receive diversity is implemented similarly, by plugging another ADC to the connector. The RSP already supports receive diversity through two distinct input buses. Moreover, once a suitable RF front end becomes available, predistortion techniques could be used to compensate for component imperfections in the path from baseband to antenna.

Although the transceiver has been designed as an R&D platform, keeping cost as low as possible was also an issue. A single transceiver prototype consisting of a DSK board and a tntSDR daughter-board costs about € 1.000, which is significantly less than any commercial product offering the same set of features. This pricing allows manufacturing a number of transceivers, which is particularly important for ad hoc network testbeds. A lot of functions, both in C++ and VHDL, can be implemented, tested and easily reused in other environments.

In particular, future versions of the demonstrator are likely not to be based on a DSK. Instead, a suitable FPGA with a soft processor or embedded RISC core is intended to replace the DSK [129]. This will provide a further enhancement and cost reduction path. The C++ source code developed for the DSP is very portable and thus, can also run in a Virtex II Pro's embedded PowerPC core, for instance. Moreover, the VHDL language has been especially designed to facilitate reuse of architectures and entities. Hence, once developed digital designs can be implemented on the Spartan II FPGAs currently employed by tntSDR, or on any other FPGA.

CHAPTER 10

Simulator/Demonstrator Integration

In parts II and III an innovative software simulator and a new hardware demonstrator have been presented, respectively. Although each of them is justified and useful as such, they are even more valuable in conjunction. When taken together, they provide an integrated environment for R&D in all seven layers of the ISO/OSI reference model. They complement each other, and allow to shorten the time span required to develop a complete, cross-layer optimized network device.

10.1 The Simulator: An Environment for L2+ R&D

In chapter 6, a novel simulator has been introduced, which is particularly useful for the simulation of L2 – L7 protocols and applications. It can be used to verify and optimize protocols intended for large scale wireless ad hoc networks, due to its global all-embracing knowledge. To a well-designed protocol, it must make no difference on which platform it actually operates – may it be a virtual station inside the simulator, the hardware demonstrator presented in the previous chapter, or any other device. As a result, the simulator serves as a development environment for protocols in L2 and above. Even some functions at the upper edge of L1 – i.e. those that are intended to run on a DSP and hence are written in C++ – may be verified in the framework provided by the simulator.

The noise accumulating PMD transceiver together with the free-space wireless medium model is an appropriate vehicle to simulate large scale wireless networks. Albeit possible, it would not be very meaningful to use a sophisticated ray-tracing engine to accurately simulate the effects of reflections, refractions, DOPPLER shifts, and so on. Of course, such high-fidelity simulations provide valuable information when designing cellular networks. What makes sense in the cellular world, must not necessarily be useful in the context of ad hoc mesh networks. Obviously boundary conditions are significantly different: in cellular networks BSs are immobile and environmental conditions are well-known. It is against the nature of unmanaged ad hoc networks to make such assumptions here. When simulating ad hoc networks, a sophisticated bit-accurate transceiver model would result in nothing but unfruitful computational overhead.

10.2   The Demonstrator: A Complement for L1 R&D

Naturally, the simulator is not intended as an environment for the development of signal processing algorithms. This gap is filled by the hardware demonstrator. It provides all means necessary to
develop, verify and optimize digital designs and embedded software to be used in actual network
devices. Functions like data whitening, interleaving, convolutional coding, constellation mapping,
etc. are implemented in the demonstrator, either as C++ classes or hardware descriptions (Ver-
ilog, VHDL, ...). The demonstrator is also able to host the L2 – L7 protocols developed in the
framework of the simulator. But here, it is difficult to evaluate performance of the protocols in
a global view. A number of devices can be set up to form a testbed, but it would not be easy to
track protocol activity in the same detail as can be done in simulation.

Recently, hardware-/software co-design has become increasingly important. Embedded systems for reactive real-time applications are implemented as mixed software-hardware systems.
Generally, software is used for features and flexibility, while hardware is used for performance.
Design of embedded systems can be subject to many different types of constraints, including
timing, size, weight, power consumption, reliability, and cost. Today, hardware and software are
designed separately. A specification is developed and sent to the hardware and software engineers.
Hardware/software partition is decided **a priori** and is adhered to as much as is possible. Any
changes in this partition may necessitate extensive, costly redesign. The present hardware demonstrator allows to shift functions between hardware and software environments. The repartitioning
can occur virtually at any time during system development.

10.3   Integrating Both Elements

Two concepts ensure that protocols implemented in WWANS are easily ported to the demonstrator platform: abstraction and separation.

10.3.1   Abstraction

At some point, certain services rely on platform specific functions. Consider a simple timer, which
is required by many different protocols as a basic service. Obviously, a timer is implemented
differently in the simulator, than in the hardware demonstrator. Even different revisions of the
hardware platform may necessitate different implementations of a timer service. Abstraction hides
the specifics of any particular platform from service users based on an agreed-upon interface. It
has been stressed that the use of C++ is something that simulator and demonstrator have in
common. Consequently, it is very natural to utilize an approved object oriented design practice
to actually implement the glue between simulator and demonstrator. Consider the following class
declaration:

```
class CTimerService
{
    // Overridables
    public:
        // ...

        // Returns the current time
        virtual double GetTime() const = 0;

        // ...
};
```

Obviously, this declares the class CTimerService. Inside the class, a pure virtual function is declared, GetTime(), which returns the current time. As the method is pure virtual, it does not have an implementation. In other words, it cannot actually be invoked. Instead, it is meant as an interface only. This can be seen as a contract between service provider and service user. A service user would make use of the class in the following way, for example:

```
void CServiceUser::UseTheTimer(CTimerService *pTimerService)
{
    const double dblCurrentTime = pTimerService->GetTime();

    // ...
}
```

While this method invocation cannot be used with an instance of the CTimerService base class, it can be used with derived, specialized classes that actually implement the functionality. A service user is able to utilize the timer service, no matter how it has actually been implemented. Inheritance is a neat feature of object oriented design. The simulator implements a timer service based on the interface defined by the CTimerService base class. The simulator's implementation is called CSimulationTimerService, and is derived from the original timer service class. Consider the following declaration:

```
class CSimulationTimerService : public CTimerService
{
    // Overrides
    public:
        // ...

        // Returns the current simulator time
        virtual double GetTime() const;

        // ...
};
```

It can be seen that the GetTime() method is overwritten. The implementation of this method (which is not shown), simply returns the simulator's DE kernel time. It can be easily imagined, what another class CDemonstratorTimerService does. This methodology is essentially the same that modern operating systems use to hide hardware specifics from applications. The corresponding part of an operating system is frequently called hardware abstraction layer (HAL). Device drivers are used to map abstract hardware functions to real devices.

10.3.2 Separation

Separation is another important principle used to obtain an integrated environment. First of all, functions that are distinctive to simulation only, need to be separated from the core functionality. In particular, events that control simulation flow and directives that are used to create such events are generally declared and defined in different source code files than the core protocols. Obviously, directives and events are not (and need not be) portable.

Second, protocols as such must be separated and interfaces between these models must be designed appropriately. Just consider IEEE 802.11 as an example. In some simulators the different parts and modules that make up this standard are merged into a single source file. While this may be appropriate for simulation alone, it renders the source code unusable for actual network devices.

10.4   Benefits of the Integrated Approach

At least two benefits of this integrated approach are easily identified. A first important aspect is reduced time-to-market. Protocols that have been thoroughly tested and revised in simulation must not be rewritten in a different programming language. Besides the intrinsic savings, this avoids a lot of potential flaws and bugs introduced by a complete rewrite. Moreover, it forces protocol designers to think about interfaces properly. Consequently, usability of the resulting protocols can be expected to be better than without such clear "border lines". This is also the reason, why standard compliance is so important: protocols that build on widely deployed standards are more likely to gain acceptance than ivory-tower solutions. Moreover, a modular structure facilitates extensibility and maintenance.

CHAPTER 11

Conclusion

This work started off with a review of state-of-the-art concepts and building blocks often faced in the context of wireless multihop ad hoc networks. Wireless network architectures in general, and the ad hoc networking paradigm in particular have been surveyed. Advantages and challenges of these "pure wireless" networks have been outlined. Fundamentals of QoS and the related terminology have been recalled.

Chapter 2 was dedicated to the design of a new architecture enabling high-performance, QoS-aware switching in multihop ad hoc networks. Founded on an in-depth analysis of the short-comings of current best-practice architectures, the innovative idea of a hybrid L2 cut-through/cell switching has been pointed out. Besides setting the scene for the elementary building blocks that have been introduced in the algorithmic part of this work, the proposed architecture presents an innovation as such.

The main algorithmic contributions of this work have been in the focus of part I. Here, enhancements to the standard IEEE 802.11 DCF have been presented. Efficient multicasts with improved reliability and methods for both, avoiding and coping with congestion, have turned out to be particularly useful as basic elements in a broad range of management protocols. Hence, these have been presented as a self-contained topic in chapter 3.

An innovative approach to distributed dynamic channel assignment has been introduced in chapter 4 for the first time: the transaction-based soft-decision radio resource management protocol, or TBSD-RRM in short. A suite of modular, ready-to-use protocols (NDP + TBSD-RRM) has been designed and simulated under realistic conditions. The novel TBSD-RRM is able to deal with the pitfalls of a shared random access channel used to exchange management frames. This key contribution offers conflict-free channel assignments for arbitrary wireless networks thanks to the underlying flexible constraint-set framework. The new idea of soft-decision table fusion under structural constraints led to fast convergence. In spite of TBSD-RRM being a distributed proto-col, it has been shown to achieve a performance comparable to that of the centralized UxDMA for a number of representative topologies. It should be stressed that RAMANATHAN's centralized al-gorithm itself presented a major improvement over prior dynamic channel assignment algorithms. In contrast to earlier proposals, TBSD-RRM does neither require complete network topology knowledge, nor multiple distributed passes over the entire network. Together with its ability of incremental operation, which is due to the stateful transaction-based architecture, TBSD-RRM has been shown to be an ideal candidate for MANETs.

In chapter 5, a further key contribution has been presented. Here, WSNs have been identified as a new field of application for a classic, powerful tool of stochastic signal processing: the KALMAN filter. A new way of reliable, yet efficient transmission of sensor values over arbitrary wireless (and wired) networks has been pointed out. Incorporating an asynchronous KALMAN filter into existing data flows results in a gain in network capacity, and consequently a reduction of power-consumption. Theoretical considerations have been supported by numerical simulations so as to prove the filter's exceptional performance in the proposed application. Moreover, this kind of filter has been discovered as a very natural choice, when it comes to implementing the advanced concept of data fusion in WSNs. In a more abstract view, the proposed methodology can be seen as an L7 approach to maintaining QoS in a hostile environment, at least for a particular class of applications.

Part II was dedicated to a new set of software tools: the Wireless Wide Area Network Simulator. All simulation results, network graphs, etc. presented throughout this work have been obtained by WWANS. Its flexible and modular architecture supports system level simulations, cross-layer optimization approaches, and further advanced research activities. The main simulator engine alone consists of 900 KB of compact hand-crafted C++ source code. At the core of WWANS, a DE kernel executes arbitrary events that control simulation flow. A very fast calendar queue implementation has been developed to manage the potentially huge event set. A suite of reusable protocols has been designed particularly for WWANS and the hardware demonstrator platform. This suite comprises IEEE 802.11 DCF, AODV, IPv4, ARP, UDP, WSDP, TBSD-RRM, NDP, WSDP client & server applications, a CBR source and a generic UDP sink. In addition, a powerful wireless medium model and a noise accumulating PMD model are dedicated to the simulator alone. Some typical flows have shown the wealth of information that can be drawn from simulations conducted with WWANS. The presented simulator output fragments have indicated the accuracy provided by each model and protocol implementation. Moreover, software tools for graphical representation of graphs, generation of simulation scripts describing random topologies, generation of channel acquisition requests, and evaluation of TBSD-RRM assignment results have been presented.

Finally, part III sketched the hardware demonstrator platform, which has also been developed in the course of this work. The initial motivation was to demonstrate that wide-band digital signal processing can make transceivers for multi-channel ad hoc networks affordable. Anyway, it soon turned out that the demonstrator, when combined with the simulator, provided a unique integrated R&D platform for future wireless mesh networks. Thanks to its flexibility, this platform allows deciding which functions to implement as digital hardware designs, and which ones as software running on general purpose DSPs or RISC processors. Moreover, this hardware/software co-design approach allows shifting functions from one world to the other even at late times during system design. This a posteriori shift causes rather little cost. Once a system specification has been completed, the previously developed and thoroughly tested C++ and VHDL components can be reused for a system-on-chip (SoC) design with only little to none modifications. Moreover, knowledge of underlying hardware architectures avoids ivory-tower solutions without practical applicability.

APPENDIX A

Reports Generated by tntSDR Diagnose Tools

A.1 Verification of TSP Programming

In this section, two typical reports are listed, which are meant to give an idea of the TSP's and RSP's features. These reports have been created with two tools particularly developed for the tntSDR platform. Both tools turn a raw memory image into a human readable textual representation and enable diagnosis by plausibility checking, etc. Contents of common function registers and the output related to the first channel (A) are listed. The following fragment shows a textual report generated by the diagnose tool ad6623chk, which has been developed in conjunction with the tntSDR hardware demonstrator platform.

```
----------------------------------------------------------------------
--
-- This report has been created by ad6623chk Version 1.0
-- Copyright(C) 2003, 2004 University of Wuppertal, Communication Theory
--
-- Report file: tsp.txt
-- Memory image: ad6623dsk.raw
-- AD6623 master clock frequency: 80 MHz
--
----------------------------------------------------------------------


---------------------------- MEMORY DUMP ----------------------------
0000: 00000093 00000000 00000000 00000000

----------------------------------------------------------------------
---------------------- COMMON FUNCTION REGISTERS ----------------------
----------------------------------------------------------------------

Clip wideband I/O...................... ENABLED
Offset binary outputs.................. ENABLED
Dual output............................ disabled
Wideband input......................... disabled
AD6623 Extensions...................... ENABLED

Channel A sync0 pin.................... disabled
Channel B sync0 pin.................... disabled
Channel C sync0 pin.................... disabled
Channel D sync0 pin.................... disabled

Start on pin sync...................... disabled
Hop on pin sync........................ disabled
Beam on pin sync....................... disabled
First sync only........................ disabled

BIST counter........................... 000000
BIST value............................. 0000

---------------------------- MEMORY DUMP ----------------------------
0100: 00000002 00000001 00000000 00000000 00000000 00000000 0000000c 00000000
0108: 00000000 00000007 00000017 00000000 00000003 00000000 0000987c 00000002
0110: 0000939e 0000f6b7 00007ce1 000069de 00046bd 000018d7 00000000 0000003f
0118: 0000001f 0000001f 0000001f 0000001f 0000001f 0000001f 0000001f 0000001f
0120: 00000000 00000000 00000000 00000000 00000000 00000000 00000000 00000000
```

169

```
0128: 00000000 00000000 00000000 00000000 00000000 00000000 00000000 00000000
0130: 00000000 00000000 00000000 00000000 00000000 00000000 00000000 00000000
0138: 00000000 00000000 00000000 00000000 00000000 00000000 00000000 00000000
0140: 00000000 00000104 00000208 0000030c 00000410 00000514 00000618 0000071c
0148: 00000820 00000924 00000a28 00000b2c 00000c30 00000d34 00000e38 00000f3c
0150: 00001040 00001144 00001248 0000134c 00001450 00001555 00001659 0000175d
0158: 00001861 00001965 00001a69 00001b6d 00001c71 00001d75 00001e79 00001f7d
0160: 00002081 00002185 00002289 0000238d 00002491 00002595 00002699 0000279d
0168: 000028a1 000029a5 00002aaa 00002bae 00002cb2 00002db6 00002eba 00002fbe
0170: 000030c2 000031c6 000032ca 000033ce 000034d2 000035d6 000036da 000037de
0178: 000038e2 000039e6 00003aea 00003bee 00003cf2 00003df6 00003efa 00003fff
0180: 0000ff4b 0000ff9b 000060e3 00000974 0000fcd0 0000ee93 0000386e 000006c1
0188: 0000fbc4 0000eb0a 00000634 0000ff2f 0000ff2f 00000634 0000eb0a 0000fbc4
0190: 000006c1 0000386e 0000ee93 0000fcd0 00000974 000060e3 0000ff9b 0000ff4b
0198: 00008555 0000cfd0 0000bd1d 0000b332 000098f3 00007fe2 0000dd45 0000bc1e
01a0: 00007c50 0000a250 00002e64 00002ed1 0000629b 0000087be 00004329 0000f7f0
01a8: 000077d9 00009baf 0000e957 0000d6c4 0000cd48 0000a61e 00005831 0000cdf6
01b0: 0000fcd8 000054e6 0000c6c0 00002bfb 000093e9 0000b7ff 0000bd4c 00008e55
01b8: 00004860 0000b337 0000cdfd 00005acc 0000ff24 000078fc 0000bd60 0000e065
01c0: 0000ff7f 0000e06e 00005fad 0000f466 0000ea3d 000093cc 00002c31 0000a3b2
01c8: 0000b667 000076f5 0000de70 0000ec07 0000a78d 000092b1 0000bde8 0000b467
01d0: 0000f8ec 0000c30f 0000ce19 00006106 0000b64c 0000e875 00006527 0000b2cc
01d8: 0000273f 0000316d 00006329 00008e5f 000095d5 0000154f 00002b08 0000eb3e
01e0: 0000e9ea 00001d2d 000054ce 0000154d 000058ef 0000691f 00006065 0000bd82
01e8: 000072ca 0000e74b 000053d2 0000de8e 00004d38 0000487e 00001fae 0000b7ed
01f0: 0000f75e 0000cbac 0000cee0 00001d40 00001f42 00003191 00008fbc 00005e99
01f8: 00003a3d 00002991 0000175f 0000e9fb 0000d6fe 0000c33f 0000f988 00007d0e


--------------------------------- MEMORY DUMP ---------------------------------
0900: 0000ff4b 0000ff9b 000060e3 00000974 0000fcd0 0000ee93 0000386e 000006c1
0908: 0000fbc4 0000eb0a 00000634 0000ff2f 0000ff2f 00000634 0000eb0a 0000fbc4
0910: 000006c1 0000386e 0000ee93 0000fcd0 00000974 000060e3 0000ff9b 0000ff4b
0918: 0000c550 0000d51 0000949c 0000ab12 0000d6d1 00005ac2 0000d444 0000bc1b
0920: 00001910 00005280 00002f70 00006cdb 0000d28e 00008794 0000634b 0000d0f0
0928: 000074d1 00009f27 00004c47 0000d6c0 0000cc58 0000261a 00007471 0000cdf6
0930: 000094f9 00000ce0 0000d7cd 00001e77 000091e9 0000056d 0000994c 0000cd44
0938: 0000796c 00007271 0000e99d 000098fc 00005f24 000059dc 0000ad0d 0000e46d
0940: 0000ff7f 0000c01a 00001fae 0000a400 0000ea3d 00007ec 00006c20 00009330
0948: 00002227 000032f5 00005e70 00008c07 0000b31f 000093b1 0000bff8 00009747
0950: 000078e8 0000581c 0000ce09 0000f107 00007656 0000e864 0000e045 000030cc
0958: 00000f3f 00003047 00006283 000088c8 00008dd5 000015cf 00003f14 0000ea7c
0960: 000011ca 0000910d 00004cce 00009f5d 0000c46c 00007e16 0000e562 00005a98
0968: 0000a2e0 0000624b 00005643 0000da8e 0000451b 00008b8 00003f8d 0000b6dd
0970: 00008716 000098bc 0000afe8 00000f14 0000d4e 000010b2 00002b59 0005a98
0978: 00001b98 0000c190 0000a7df 000069d7 000006d4 00009bb 0000f989 0000759e
0980: 000015c0 000029bd 00001dc7 00002439 0000aed0 0000b334 0000e704 0000b33c
0988: 0000c41f 0000ed41 00006d15 00004c6f 0000568d 00004180 0000b171 0000abea
0990: 0000bcda 000011cd 00003bd5 0000d736 0000a2cb 00004004 000061cc 00089fc
0998: 00006f4d 00005095 0000d717 0000a477 0000be25 0000b847 0000f741 0000eaf2
09a0: 00003204 000031cd 0000336e 0000a591 0000731a 0000952c 0000d9df 000055a6
09a8: 00000616 00000d5b 00004974 0001ce0 0000dcdc 0000ae4d 00002928 0000869d
09b0: 0000a77c 00002bc0 0000772c 0000351d 0000dffe 000096e6 00005de8 00005ed0
09b8: 00009c5d 0000d8da 0000c721 00000414 000044f9 00005fb5 0000bc58 0000cd0b
09c0: 0000544e 0000cf04 0000a175 00003f08 00008d22 00002634 0000070c 00005134
09c8: 0000f18c 0000f3d1 000013ea 00003543 0000fce7 000021c3 0000c523 0000b0c1
09d0: 00006547 00008590 0000472e 0000c135 00005f83 00005691 000075d6 0000957a
09d8: 00007ec8 0000b611 0000b59e 0000c7d 0000f8ab 0000664b 00004926 000094d1
09e0: 00004bd9 00001298 0000e6c5 0000496f 0000e72d 00004ca2 0000cbdf 0000de89
09e8: 0000999c 000001d1 0000a51b 0000771a 00001155 0000dd70 00007378 00002e45
09f0: 00005b9b 0000a5ec 00001f07 00000458 00004579 0000a955 00005b82 00004207
09f8: 00003e3b 0000d434 00004555 00009aa7 0000665d 000085d3 0000c034 00000c4f
```

```
------------------------------------------------------------------------------
---------------------- CHANNEL FUNCTION REGISTERS (A) ------------------------
------------------------------------------------------------------------------


Start hold-off counter.................. 0002
Start sync select........................ sync0

NCO scale................................ -12 dB
NCO clear phase accumulator on sync..... disabled
NCO phase dither......................... disabled
NCO amplitude dither..................... disabled
NCO frequency value...................... 00000000 (0.000000 MHz)
NCO frequency update hold-off........... 0000
NCO hop sync select...................... sync0
NCO phase offset......................... 0000 °(0.000000)
NCO phase offset update hold-off........ 0000
NCO phase (beam) sync select............ sync0

CIC scale................................ 0c
CIC2 decimation (M2-1)................... 000
CIC2 interpolation (L2-1)............... 000
CIC5 interpolation (L5-1)............... 07

RCF taps A............................... 17
RCF taps B............................... 00
RCF coefficient offset................... 00
RCF taps per phase - 1.................. 3
RCF mode................................. FIR
RCF pseudo-random sequence length....... short (15)
RCF pseudo-random input................. disabled
RCF compact FIR word length............. 16 bits (8i + 8q)
RCF serial clock divisor................ 00
RCF all-pass phase equalizer............ disabled
RCF coarse scale........................ 0 dB
RCF fine scale factor................... 261f
```

```
RCF scale hold-off..................... 0002
RCF time slot sync select.............. sync0
RCF phase equalizer coefficient 1....... 939e
RCF phase equalizer coefficient 2....... f6b7
RCF FIR-PSK magnitude 0................. 7ce1
RCF FIR-PSK magnitude 1................. 69de
RCF FIR-PSK magnitude 2................. 46bd
RCF FIR-PSK magnitude 3................. 18d7

SCLK mode.............................. master
SDFO mode.............................. frame request
SDFI mode.............................. internal frame request
Serial time slot sync.................. disabled

Ramping................................ disabled
Ramp interpolation..................... disabled
Ramp length, mode 0 (R0-1)............. 3f
Ramp length, mode 1 (R1-1)............. 1f
Ramp rest time, Q...................... 00

------------------- CHANNEL SETTINGS OVERVIEW (A) ---------------------

RCF coefficient count.................. 24
RCF taps per phase..................... 4
RCF interpolation...................... 6
RCF fine scale......................... 1.19128 (1.52031 dB)
RCF coarse scale....................... 1.00000 (0.000000 dB)
CIC5 interpolation..................... 8
CIC2 interpolation..................... 1
CIC2 decimation........................ 1
Recommended CIC scale.................. 12
Total channel interpolation............ 48
Channel input rate (SDFO frequency).... 1666.67 kSPS
SCLK cycles required per sample........ 33
Minimum SCLK frequency................. 55.0000 MHz
Master mode SCLK frequency............. 80.0000 MHz

--------------------- SANITY CHECK LIST RESULTS (A) ----------------------

RCF mode 0. Taps B zero?............... pass.
RCF interpolation integer?............. pass.
RCF interpolation <= 8?................ pass.
Timing constraint met?................. pass.
Data memory constraint met?............ pass.
Taps per phase constraint 1 met?....... pass.
Taps per phase constraint 2 met?....... pass.
CIC scale in range 4 - 32?............. pass.
CIC scale appropriate?................. pass.
CIC5 interpolation <= 32?.............. pass.
CIC2 interpolation appropriate?........ pass.
CIC2 decimation appropriate?........... pass.
CIC inequality met?.................... pass.
SCLK mode appropriate?................. pass.
SCLK divider appropriate?.............. pass.

------------------------- RCF COEFFICIENT VALUES -------------------------

Coefficient h(0)....................... -0.00552368
Coefficient h(1)....................... -0.0249023
Coefficient h(2)....................... -0.0330811
Coefficient h(3)....................... -0.00637817
Coefficient h(4)....................... 0.0527649
Coefficient h(5)....................... 0.0738525
Coefficient h(6)....................... -0.00308228
Coefficient h(7)....................... -0.136139
Coefficient h(8)....................... -0.163757
Coefficient h(9)....................... 0.0484619
Coefficient h(10)...................... 0.440857
Coefficient h(11)...................... 0.756927
Coefficient h(12)...................... 0.756927
Coefficient h(13)...................... 0.440857
Coefficient h(14)...................... 0.0484619
Coefficient h(15)...................... -0.163757
Coefficient h(16)...................... -0.136139
Coefficient h(17)...................... -0.00308228
Coefficient h(18)...................... 0.0738525
Coefficient h(19)...................... 0.0527649
Coefficient h(20)...................... -0.00637817
Coefficient h(21)...................... -0.0330811
Coefficient h(22)...................... -0.0249023
Coefficient h(23)...................... -0.00552368

------------------------- ANALYSIS OF COEFFICIENTS -------------------------

Phase 0, center gain................... 0.822174 (-1.70072 dB)
Phase 1, center gain................... 0.332581 (-9.56206 dB)
Phase 2, center gain................... -0.154755 (-1.#IND0 dB)
Phase 3, center gain................... -0.154755 (-1.#IND0 dB)
Phase 4, center gain................... 0.332581 (-9.56206 dB)
Phase 5, center gain................... 0.822174 (-1.70072 dB)

Phase 0, worst-case peak............... 0.839386
Phase 1, worst-case peak............... 0.654663
Phase 2, worst-case peak............... 0.251678
Phase 3, worst-case peak............... 0.251678
Phase 4, worst-case peak............... 0.654663
Phase 5, worst-case peak............... 0.839386
```

```
----------------------- SCALING ANALYSIS -----------------------
RCF coarse scaler gain................. 1.00000  (0.000000 dB)
RCF fine scaler gain................... 1.19128  (1.52031 dB)
CIC scaler gain........................ 0.000244141 (-72.2472 dB)
CIC5 interpolation gain................ 4096.00  (72.2472 dB)
CIC2 interpolation gain................ 1.00000  (0.000000 dB)
CIC2 decimation gain................... 1.00000  (0.000000 dB)
NCO gain............................... 0.250000  (-12.0412 dB)

Max. worst-case peak @ RCF out......... 0.839386
Max. peak @ RCF coarse scaler out...... 0.839386
Max. peak @ RCF fine scaler out........ 0.999947
Max. peak @ CIC scaler out............. 0.000244128
Max. peak @ CIC5 out................... 0.999947
Max. peak @ CIC2i out.................. 0.999947
Max. peak @ CIC2d out.................. 0.999947
Max. peak @ NCO out.................... 0.249987
```

Listing A.1: Report for the TSP

A.2   Verification of RSP Programming

A similar tool like ad6623chk has been developed for the RSP. The following output has been created by ad6624chk:

```
-----------------------------------------------------------------
--
-- This report has been created by ad6624chk Version 1.0
-- Copyright(C) 2004 University of Wuppertal, Communication Theory
--
-- Report file: rsp.txt
-- Memory image: ad6624dsk.raw
-- AD6624 master clock frequency: 80 MHz
--
-----------------------------------------------------------------


----------------------- MEMORY DUMP -----------------------
0000: 00000000 000001ff 0000001e 00000000 00000000 000001ff 0000001e 00000000


-------------------- INPUT PORT CONTROL REGISTERS --------------------

Port A, lower threshold................ 000
Port A, upper threshold................ 1ff
Port A, dwell time..................... 0001e
Port A, output polarity................ disabled
Port A, interleaved channels........... disabled
Port A, linearization hold-off......... 0

Port B, lower threshold................ 000
Port B, upper threshold................ 1ff
Port B, dwell time..................... 0001e
Port B, output polarity................ disabled
Port B, interleaved channels........... disabled
Port B, linearization hold-off......... 0

----------------------- MEMORY DUMP -----------------------
0100: 000fdaf3 000fe05f 0000347b 0000a11a 000046f7 000f00cc 000e3fef 000009e8
0108: 000447df 0007fffe 0007fffe 000447df 000009e8 000e3fef 000f00cc 000046f7
0110: 0000a11a 0000347b 000fe05f 000fdaf3 00000000 00000000 00000000 00000000
0118: 00000000 00000000 00000000 00000000 00000000 00000000 00000000 00000000
0120: 00000000 00000000 00000000 00000000 00000000 00000000 00000000 00000000
0128: 00000000 00000000 00000000 00000000 00000000 00000000 00000000 00000000
0130: 00000000 00000000 00000000 00000000 00000000 00000000 00000000 00000000
0138: 00000000 00000000 00000000 00000000 00000000 00000000 00000000 00000000
0140: 00000000 00000000 00000000 00000000 00000000 00000000 00000000 00000000
0148: 00000000 00000000 00000000 00000000 00000000 00000000 00000000 00000000
0150: 00000000 00000000 00000000 00000000 00000000 00000000 00000000 00000000
0158: 00000000 00000000 00000000 00000000 00000000 00000000 00000000 00000000
0160: 00000000 00000000 00000000 00000000 00000000 00000000 00000000 00000000
0168: 00000000 00000000 00000000 00000000 00000000 00000000 00000000 00000000
0170: 00000000 00000000 00000000 00000000 00000000 00000000 00000000 00000000
0178: 00000000 00000000 00000000 00000000 00000000 00000000 00000000 00000000
0180: 00000001 00000000 00000000 00000001 00000000 00000000 00000000 00000000
0188: 00000010 00000010 00000000 00000010 00000011 00000010 00000010 00000010
0190: 00000007 00000000 000000c6 00000000 00000001 00000000 00000000 00000000
0198: 00000007 00000000 000000c6 00000000 00000001 00000000 00000000 00000000
01a0: 00000002 00000000 00000013 00000000 00000003 000468e 00000000 00000000
01a8: 00000000 00000030 00000013 00000000 00000003 000468e 00000000 00000000
01b0: 00000000 00000000 00000000 00000000 00000000 00000000 00000000 00000000
01b8: 00000000 00000000 00000000 00000000 00000000 00000000 00000000 00000000
01c0: 00000000 00000000 00000000 00000000 00000000 00000000 00000000 00000000
01c8: 00000000 00000000 00000000 00000000 00000000 00000000 00000000 00000000
```

```
01d0: 00000000 00000000 00000000 00000000 00000000 00000000 00000000 00000000
01d8: 00000000 00000000 00000000 00000000 00000000 00000000 00000000 00000000
01e0: 00000000 00000000 00000000 00000000 00000000 00000000 00000000 00000000
01e8: 00000000 00000000 00000000 00000000 00000000 00000000 00000000 00000000
01f0: 00000000 00000000 00000000 00000000 00000000 00000000 00000000 00000000
01f8: 00000000 00000000 00000000 00000000 00000000 00000000 00000000 00000000


------------------------ MEMORY DUMP ----------------------------
0900: 00000000 00000000 00000000 00000000 00000000 00000000 00000000 00000000
0908: 00000000 00000000 00000000 00000000 00000000 00000000 00000000 00000000
0910: 00000000 00000000 00000000 00000000 00000000 00000000 00000000 00000000
0918: 00000000 00000000 00000000 00000000 00000000 00000000 00000000 00000000
0920: 00000000 00000000 00000000 00000000 00000000 00000000 00000000 00000000
0928: 00000000 00000000 00000000 00000000 00000000 00000000 00000000 00000000
0930: 00000000 00000000 00000000 00000000 00000000 00000000 00000000 00000000
0938: 00000000 00000000 00000000 00000000 00000000 00000000 00000000 00000000
0940: 00000000 00000000 00000000 00000000 00000000 00000000 00000000 00000000
0948: 00000000 00000000 00000000 00000000 00000000 00000000 00000000 00000000
0950: 00000000 00000000 00000000 00000000 00000000 00000000 00000000 00000000
0958: 00000000 00000000 00000000 00000000 00000000 00000000 00000000 00000000
0960: 00000000 00000000 00000000 00000000 00000000 00000000 00000000 00000000
0968: 00000000 00000000 00000000 00000000 00000000 00000000 00000000 00000000
0970: 00000000 00000000 00000000 00000000 00000000 00000000 00000000 00000000
0978: 00000000 00000000 00000000 00000000 00000000 00000000 00000000 00000000


------------------------------------------------------------------------
------------------ CHANNEL FUNCTION REGISTERS (A) ----------------------
------------------------------------------------------------------------


Channel sleep.......................... ENABLED

Soft sync start........................ disabled
Soft sync hop.......................... disabled

Pin sync start......................... disabled
Pin sync hop........................... disabled
Pin sync first only.................... disabled

Start hold-off counter................. 0001

NCO frequency update hold-off.......... 0000
NCO frequency register 0............... 0000
NCO frequency register 1............... 0000
NCO frequency value.................... 00000000 (0.000000 MHz)
NCO phase offset....................... 0000 °(0.000000)
NCO phase dither....................... disabled
NCO amplitude dither................... disabled
NCO clear phase accumulator on hop..... disabled

Bypass (A -> I, B -> Q)................ disabled
Wideband input select.................. Port A
Sync Input Select...................... SYNCA

CIC2 decimation (M2-1)................. 007
CIC2 interpolation (L2-1).............. 000
CIC2 scale, loud....................... 06
CIC2 scale, quiet...................... 06
CIC2 scale, exponent weight............ disabled
CIC2 scale, exponent invert............ disabled

CIC5 decimation (M5-1)................. 01
CIC5 scale............................. 00

RCF decimation (M-1)................... 02
RCF decimation phase................... 00
RCF taps - 1........................... 13
RCF coefficient offset................. 00
RCF output format...................... fixed point
RCF output scale....................... 3
RCF force output scale................. disabled
RCF use common exponent................ disabled
RCF memory bank 1...................... disabled
RCF input select....................... own
RCF bypass BIST........................ disabled

BIST signature, I...................... 468e
BIST signature, Q...................... 0000
BIST, #outputs to accumulate........... 00000
BIST in progress....................... no
BIST, CMEM result...................... pass
BIST, DMEM result...................... pass

Serial clock divisor................... 0
Serial bus master...................... ENABLED
Serial output word length.............. 16 bit
SDFS mode.............................. single pulse
Map RCF data to BIST registers......... disabled

------------------- RCF COEFFICIENT VALUES -------------------

Coefficient h(0)....................... -0.0180912
Coefficient h(1)....................... -0.0154438
Coefficient h(2)....................... 0.0256252
Coefficient h(3)....................... 0.0786629
Coefficient h(4)....................... 0.0346508
Coefficient h(5)....................... -0.124611
Coefficient h(6)....................... -0.218782
```

```
Coefficient h(7)......................... 0.00483704
Coefficient h(8)......................... 0.535093
Coefficient h(9)......................... 0.999996
Coefficient h(10)........................ 0.999996
Coefficient h(11)........................ 0.535093
Coefficient h(12)........................ 0.00483704
Coefficient h(13)........................ -0.218782
Coefficient h(14)........................ -0.124611
Coefficient h(15)........................ 0.0346508
Coefficient h(16)........................ 0.0786629
Coefficient h(17)........................ 0.0256252
Coefficient h(18)........................ -0.0154438
Coefficient h(19)........................ -0.0180912


-------------------- CHANNEL SETTINGS OVERVIEW (A) --------------------

CIC2 decimation.......................... 8
CIC2 interpolation....................... 1
CIC2 scale (recommendation).............. 6
CIC2 scale, loud (actual)................ 6
CIC2 scale, quiet (actual)............... 6
CIC5 decimation.......................... 2
CIC5 scale (recommendation).............. 0
CIC5 scale (actual)...................... 0
RCF decimation........................... 3
RCF coefficient count.................... 20
Total channel decimation................. 48
Channel output rate (SDFO frequency).... 1666.67 kSPS
SCLK cycles required per sample......... 33
Minimum SCLK frequency................... 55.0000 MHz
Master mode SCLK frequency............... 80.0000 MHz


---------------------- SANITY CHECK LIST RESULTS (A) ----------------------

CIC2 rate change <= 1?................... pass.
CIC2 scale appropriate?................. pass.
CIC5 decimation in range 2 to 32?....... pass.
CIC5 scale appropriate?................. pass.
RCF decimation <= 32?................... pass.
RCF data memory constraint met?......... pass.
RCF tap constraint met?................. pass.
SCLK mode appropriate?.................. pass.
SCLK divider appropriate?............... pass.

------------------------ ANALYSIS OF COEFFICIENTS ------------------------

Channel center gain...................... 2.60387 (8.31240 dB)
Worst-case peak.......................... 4.11159 <<< *OVERFLOW*

---------------------------- SCALING ANALYSIS ----------------------------

CIC2 scaler gain......................... 0.0156250 (-36.1236 dB)
CIC5 scaler gain......................... 0.0156250 (-36.1236 dB)
RCF output scaler gain.................. 1.00000 (0.000000 dB)

Input level.............................. 0.999939
Max peak @ CIC2 scaler out.............. 0.999939
Max peak @ CIC5 scaler out.............. 0.999939
Max peak @ RCF output scaler............ 0.999939
```

Listing A.2: Report for the RSP

[1] E. D. Kaplan, Ed., **Understanding GPS: Principles and Applications**, Artech House, 1996.

[2] S. M. Redl, M. K. Weber, and M. W. Oliphant, **An Introduction to GSM**, Artech House, 1995.

[3] G. Heine, **GSM Networks: Protocols, Terminology, and Implementation**, Artech House, 1999.

[4] J. P. Castro, **The UMTS Network and Radio Access Technology**, John Wiley & Sons, 2001.

[5] F. Muratore, Ed., **UMTS: Mobile Communications for the Future**, John Wiley & Sons, 2001.

[6] J. Postel, Ed., **RFC 791: Internet Protocol – DARPA Internet Program – Protocol Specification**, IETF, Sept. 1981.

[7] F. Baker, Ed., **RFC 1812: Requirements for IP Version 4 Routers**, IETF, June 1995.

[8] S. Deering and R. Hinden, **RFC 2460: Internet Protocol, Version 6 (IPv6) – Specification**, IETF, Dec. 1998.

[9] Z. Haas, J. Deng, B. Liang, P. Papadimitatos, and S. Sajama, "Wireless ad hoc networks," in **Encyclopedia of Telecommunications**, John Proakis, Ed. John Wiley & Sons, 2002.

[10] ANSI/IEEE Standard 802.11, **IEEE Standard for Information technology – Telecommunications and information exchange between systems – Local and metropolitan area networks – Specific requirements. Part 11: Wireless LAN Medium Access Control (MAC) and Physical Layer (PHY) Specifications**, 1999, ISO/IEC 8802-11:1999(E).

[11] R. S. Kahn, J. Gronemeyer, J. Burchfiel, and R. Kunzelman, "Advances in packet radio technology," **Proceedings of the IEEE**, vol. 66, no. 11, pp. 1468–1496, Nov. 1978.

[12] N. Shacham and J Westcott, "Future directions in packet radio architectures and protocols," **Proceedings of the IEEE**, vol. 75, no. 1, pp. 83–99, Jan. 1987.

[13] B. Leiner, R. Ruth, and A. R. Sastry, "Goals and challenges of the DARPA GloMo program," **IEEE Personal Communications Magazine**, vol. 3, no. 6, pp. 34–43, Dec. 1996.

[14] F. Xiong, **Digital Modulation Techniques**, Artech House, 2000.

[15] W. W. Lu, "4G mobile research in asia," **IEEE Communications Magazine**, vol. 41, no. 3, pp. 104–106, Mar. 2003, (guest editorial).

[16] W. Zirwas, T. Giebel, N. Esseling, E. Schulz, and J. Eichinger, "Broadband multi hop networks with reduced protocol overhead," in **Proceedings of European Wireless**, Florence, Italy, Feb. 2002, vol. 1, pp. 333–339.

[17] A. J. Goldsmith and S. B. Wicker, "Design challenges for energy-constrained ad hoc wireless networks," **IEEE Wireless Communications**, vol. 9, no. 4, pp. 8–27, Aug. 2002.

[18] Z. Haas and S. Tabrizi, "On some challenges and design choices in ad hoc communications," in **Proceedings of the IEEE Military Communications Conference (MILCOM)**, Boston, MA, USA, Oct. 1998.

[19] G. N. Aggélou, "An integrated platform for ad hoc GSM cellular communications," in **The Handbook of Ad hoc Wireless Networks**, M. Ilyas, Ed., chapter 10. CRC Press, 2002.

[20] M. Siebert, M. Lott, and M. Weckerle, "Enhanced radio resource management introducing smart direct link concepts," in **Proceedings of the 5th European Personal Mobile Communications Conference (EPMCC)**, Glasgow, Scotland, UK, Apr. 2003, pp. 323–327.

[21] P. Stavroulakis, Ed., **Reliability, Survivability and Quality of Large Scale Telecommunication Systems – Case Study: Olympic Games**, John Wiley & Sons, 2003.

[22] B. Bing, **High-Speed Wireless ATM and LANs**, Artech House, 2000.

[23] S. Ramanathan and M. E. Steenstrup, "Hierarchically-organized, multihop mobile wireless networks for quality-of-service support," **ACM-Baltzer Journal of Mobile Networks and Applications**, vol. 3, no. 1, pp. 101–119, June 1998.

[24] P. Mohapatra, J. Li, and C. Gui, "QoS in mobile ad hoc networks," **IEEE Wireless Communications**, vol. 10, no. 3, pp. 44–52, June 2003.

[25] International Telecommunication Union, "Series G: Transmission Systems and Media, Digital Systems and Networks – Quality of service and performance – End-user multimedia QoS categories," ITU-T Recommendation G.1010, Telecommunication Standardization Sector of ITU, Nov. 2001.

[26] O. C. Ibe, **Fixed Broadband Wireless Access Networks and Services**, John Wiley & Sons, 2002.

[27] R. Braden, L. Zhang, S. Berson, S. Herzog, and S. Jamin, Eds., **RFC 2205: Resource Reservation Protocol (RSVP) – Version 1 Functional Specification**, IETF, Sept. 1997.

[28] J. Gozdecki, A. Jajszczyk, and R. Stankiewicz, "Quality of service terminology in IP networks," **IEEE Communications Magazine**, vol. 41, no. 3, pp. 153–159, Mar. 2003.

[29] P. Lorenz, A. Jamalipour, and D. A. Khotimsky, "Quality of service in IP and wireless networks," **IEEE Communications Magazine**, vol. 42, no. 6, pp. 70–71, June 2004, (guest editorial).

[30] T. S. Rappaport, **Wireless Communications: Principles and Practice**, Prentice Hall, 2nd edition, 2002.

[31] L. Kleinrock and F. A. Tobagi, "Packet switching in radio channels: Part I – Carrier sense multiple-access modes and their throughput-delay characteristics," **IEEE Transactions on Communications**, vol. 23, no. 12, pp. 1400–1416, Dec. 1975.

[32] L. Kleinrock and F. A. Tobagi, "Packet switching in radio channels: Part II – The hidden terminal problem in carrier sense multiple-access and the busy-tone solution," **IEEE Transactions on Communications**, vol. 23, no. 12, pp. 1417–1433, Dec. 1975.

[33] ANSI/IEEE Standard 802.11a, **Supplement to IEEE Standard for Information technology – Telecommunications and information exchange between systems – Local and metropolitan area networks – Specific requirements. Part 11: Wireless LAN Medium Access Control (MAC) and Physical Layer (PHY) Specifications – High-speed Physical Layer in the 5 GHz Band**, 1999, ISO/IEC 8802-11:1999/Amd 1:2000(E).

[34] ANSI/IEEE Standard 802.11b, **Supplement to IEEE Standard for Information technology – Telecommunications and information exchange between systems – Local and metropolitan area networks – Specific requirements. Part 11: Wireless LAN Medium Access Control (MAC) and Physical Layer (PHY) Specifications – Higher-speed Physical Layer Extension in the 2.4 GHz Band**, Sept. 1999.

[35] ANSI/IEEE Standard 802.11g, **IEEE Standard for Information technology – Telecommunications and information exchange between systems – Local and metropolitan area networks – Specific requirements. Part 11: Wireless LAN Medium Access Control (MAC) and Physical Layer (PHY) Specifications – Amendment 4: Further Higher Data Rate Extension in the 2.4 GHz Band**, June 2003.

[36] ANSI/IEEE Standard 802.11h, **IEEE Standard for Information technology – Telecommunications and information exchange between systems – Local and metropolitan area networks – Specific requirements. Part 11: Wireless LAN Medium Access Control (MAC) and Physical Layer (PHY) Specifications – Amendment 5: Spectrum and Transmit Power Management Extensions in the 5 GHz band in Europe**, Oct. 2003.

[37] ANSI/IEEE Standard 802.2, **IEEE Standard for Information technology – Telecommunications and information exchange between systems – Local and metropolitan area networks – Specific requirements. Part 2: Logical Link Control**, 1998, ISO/IEC 8802-2:1998.

[38] ANSI/IEEE Standard 802.3, **IEEE Standard for Information technology - Telecommunications and information exchange between systems - Local and metropolitan area networks - Specific requirements. Part 3: Carrier sense multiple access with collision detection (CSMA/CD) access method and physical layer specifications**, Mar. 2002.

[39] IEEE Standard 802, **IEEE Standard for Local and Metropolitan Area Networks: Overview and Architecture**, Mar. 2001.

[40] S. Mangold, S. Choi, G. R. Hiertz, O. Klein, and B. Walke, "Analysis of IEEE 802.11e for QoS support in wireless LANs," **IEEE Wireless Communications**, vol. 10, no. 6, pp. 40–50, Dec. 2003.

[41] Bluetooth Special Interest Group, **Specification of the Bluetooth System, Version 1.2**, Nov. 2003, Available at http://www.bluetooth.org.

[42] IEEE Standard 802.15.1, **IEEE Standard for Information technology – Telecommunications and information exchange between systems – Local and metropolitan area networks – Specific requirements. Part 15.1: Wireless Medium Access Control (MAC) and Physical Layer (PHY) Specifications for Wireless Personal Area Networks (WPANs)**, June 2002.

[43] IEEE Standard 802.16, **IEEE Standard for Local and metropolitan area networks. Part 16: Standard Air Interface for Fixed Broadband Wireless Access Systems**, Apr. 2002.

[44] IEEE Standard 802.16a, **IEEE Standard for Local and metropolitan area networks. Part 16: Air Interface for Fixed Broadband Wireless Access Systems – Amendment 2: Medium Access Control Modifications and Additional Physical Layer Specifications for 2–11 GHz**, Apr. 2003.

[45] IEEE 802.20 Working Group, **Draft 802.20 Permanent Document – System Requirements for IEEE 802.20 Mobile Broadband Wireless Access Systems – Version 14**, July 2004.

[46] C. Perkins, E. Belding-Royer, and S. Das, **RFC 3561: Ad hoc On-Demand Distance Vector (AODV) Routing**, IETF, July 2003, (Experimental Protocol).

[47] D. Johnson and D. Maltz, "Protocols for adaptive wireless and mobile networking," **IEEE Personal Communications Magazine**, vol. 3, no. 1, pp. 34–41, Feb. 1996.

[48] C. Perkins, Ed., **RFC 3220: IP Mobility Support for IPv4**, IETF, Jan. 2002.

[49] D. A. Maltz, J. Broch, and D. B. Johnson, "Quantitative lessons from a full-scale multi-hop wireless ad hoc network testbed," in **Proceedings of the IEEE Wireless Communications and Networking Conference (WCNC)**, Chicago, IL, USA, Sept. 2000, vol. 3, pp. 992–997.

[50] K. Fall and K. Varadhan, Eds., **The ns Manual**, The VINT Project (UC Berkeley, LBL, USC/ISI, and Xerox PARC), Aug. 2000, Available at http://www.isi.edu/nsnam/ns/ns-documentation.html.

[51] B. A. Chambers, "The grid roofnet: A rooftop ad hoc wireless network," M.S. thesis, Massachusetts Institute of Technology, May 2002.

[52] D. S. J. De Couto, D. Aguayo, J. Bicket, and R. Morris, "A high-throughput path metric for multi-hop wireless routing," in **Proceedings of the 9th Annual International Conference on Mobile Computing and Networking (MobiCom)**, San Diego, CA, USA, Sept. 2003, pp. 134–146.

[53] B. Schrick and M. J. Riezenman, "Wireless broadband in a box," **IEEE Spectrum**, vol. 39, no. 6, pp. 38–42, June 2002.

[54] S. M. Cherry, "Broadband a go-go," **IEEE Spectrum**, vol. 40, no. 6, pp. 20–25, June 2003.

[55] M. A. Mayor and P. A. Gilmour, **Communications Method for a Code Division Multiple Access System without a Base Station**, United States Patent and Trademark Office, Aug. 1999, US Patent No. 5,943,322.

[56] J. Garcia-Luna-Aceves, C. Fullmer, E. Madruga, D. Beyer, and T. Frivold, "Wireless internet gateways (WINGS)," in **Proceedings of the IEEE Military Communications Conference (MILCOM)**, Monterey, CA, USA, Nov. 1997, pp. 1271–1276.

[57] E. W. Dijkstra, "A note on two problems in connexion with graphs," in **Numerische Mathematik**, 1959, vol. 1, pp. 269–271.

[58] S. Ramanathan, "On the performance of ad hoc networks with beamforming antennas," in **Proceedings of the 2nd ACM International Symposium on Mobile ad hoc Networking and Computing (MobiHoc)**, Long Beach, CA, USA, Oct. 2001, pp. 95–105.

[59] J.-P. Hubaux, J.-Y. Le Boudec, S. Giordano, M. Hamdi, L. Blazevič, L. Buttayán, and M. Vojnovič, "Towards mobile ad-hoc WANs: Terminodes," in **Proceedings of the IEEE Wireless Communications and Networking Conference (WCNC)**, Chicago, IL, USA, Sept. 2000.

[60] J.-P. Hubaux, Th. Gross, J.-Y. Le Boudec, and M. Vetterli, "Towards self-organized mobile ad hoc networks: the Terminodes project," **IEEE Communications Magazine**, vol. 39, no. 1, Jan. 2001.

[61] L. Blazevič, S. Giordano, and J.-Y. Le Boudec, "Self-organizing wide-area routing," in **Proceedings of The 4th World Multiconference on Systemics, Cybernetics and Informatics (SCI)**, Orlando, FL, USA, July 2000.

[62] S. Čapkun, M. Hamdi, and J.-P. Hubaux, "GPS-free positioning in mobile ad-hoc networks," in **Proceedings of the 34th Annual Hawaii International Conference on System Sciences**, Jan. 2001.

[63] The Wireless World Research Forum (WWRF), **Book of Vision 2001**, Available at http://www.wireless-world-research.org.

[64] Mobile IT Forum, **Flying Carpet - Towards the 4th Generation Mobile Communications Systems**, May 2003, Available at http://www.mitf.org.

[65] P. Mähönen and G. C. Polyzos, "European R&D on fourth-generation mobile and wireless IP networks," **IEEE Personal Communications Magazine**, vol. 8, no. 6, pp. 6–7, Dec. 2001, (guest editorial).

[66] B. Arroyo-Fernández, J. Fernandes, and R. Prasad, "Composite reconfigurable wireless networks: The EU R&D path toward 4G," **IEEE Communications Magazine**, vol. 42, no. 5, pp. 62–63, May 2004, (guest editorial).

[67] S. Xu and T. Saadawi, "Does the IEEE 802.11 MAC protocol work well in multihop wireless ad hoc networks?," **IEEE Communications Magazine**, vol. 39, no. 6, pp. 130–137, June 2001.

[68] J. Postel, **RFC 768: User Datagram Protocol**, IETF, Aug. 1980.

[69] J. Postel, **RFC 793: Transmission Control Protocol**, IETF, Sept. 1981.

[70] D. C. Plumer, **RFC 826: An Ethernet Address Resolution Protocol – or – Converting Network Protocol Addresses to 48 bit Ethernet Address for Transmission on Ethernet Hardware**, IETF, Nov. 1982.

[71] J. Postel and J. Reynolds, **RFC 1042: A Standard for the Transmission of IP Datagrams over IEEE 802 Networks**, IETF, Feb. 1988.

[72] W. Hoseit and H.-J. Meckelburg, **Data Transmission System with Relay Stations between a Source Station and a Target Station**, World Intellectual Property Organization, 1999, International Patent No. WO 99/12279.

[73] H.-J. Meckelburg and M. Horn, **Method for Operating an Ad-Hoc Network for the Wireless Data Transmission of Synchronous and Asynchronous Messages**, World Intellectual Property Organization, 2002, International Patent No. WO 02/49274 A2.

[74] E. Rosen, A. Viswanathan, and R. Callon, **RFC 3031: Multiprotocol Label Switching Architecture**, IETF, Jan. 2001.

[75] A. Acharya, A. Misra, and S. Bansal, "High-performance architectures for IP-based multihop 802.11 networks," **IEEE Wireless Communications**, vol. 10, no. 5, pp. 22–28, Oct. 2003.

[76] G. R. Hiertz, J. Habetha, E. Weiß, and S. Mangold, "A cut-through switching technology for IEEE 802.11," in **Proceedings of the IEEE 6th Circuits and Systems Symposium on Emerging Technologies: Frontiers of Mobile and Wireless Communication**, Shanghai, China, May 2004, vol. 2, pp. 663–666.

[77] F. H. P. Fitzek, D. Angelini, G. Mazzini, and M. Zorzi, "Design and performance of an enhanced 802.11 MAC protocol for multihop coverage extension," **IEEE Wireless Communications**, vol. 10, no. 6, pp. 30–39, Dec. 2003.

[78] A. Honarbacht and A. Kummert, "A testbed for next generation wireless multi-hop networks," in **Proceedings of the International Workshop on Multimedia Communications and Services (MCS)**, Kielce, Poland, Apr. 2003, pp. 61–65.

[79] H. Balakrishnan, C. L. Barrett, V. S. A. Kumar, M. V. Marathe, and S. Thite, "The distance-2 matching problem and its relationship to the MAC-layer capacity of ad hoc wireless networks," **IEEE Journal on Selected Areas in Communications**, vol. 22, no. 6, pp. 1069–1079, Aug. 2004.

[80] E. M. Royer and C.-K. Toh, "A review of current routing protocols for ad hoc mobile wireless networks," in **IEEE Personal Communications Magazine**, Apr. 1999.

[81] S. Ramanathan and M. E. Steenstrup, "A survey of routing techniques for mobile communications networks," **ACM-Baltzer Journal of Mobile Networks and Applications**, vol. 1, no. 2, pp. 89–104, 1996.

[82] C. R. Lin and J.-S. Liu, "QoS routing in ad hoc wireless networks," **IEEE Journal on Selected Areas in Communications**, vol. 17, no. 8, pp. 1426–1438, Aug. 1999.

[83] M. E. Steenstrup, "Dynamic multipoint virtual circuits for multimedia traffic in multihop mobile wireless networks," in **Proceedings of the IEEE Wireless Communications and Networking Conference (WCNC)**, New Orleans, Sept. 1999, vol. 2, pp. 1018–1022.

[84] G. Crippen and T. Havel, **Distance Geometry and Molecular Conformation**, John Wiley & Sons, 1988.

[85] D. Niculescu, "Positioning in ad hoc sensor networks," **IEEE Network**, vol. 18, no. 4, pp. 24–29, July 2004.

[86] A. Honarbacht and A. Kummert, "A modular radio resource management scheme for wireless networks," in **Proceedings of the International Workshop on Multimedia Communications and Services (MCS)**, Kielce, Poland, Apr. 2003, pp. 67–70.

[87] A. Honarbacht, M. Pieper, and A. Kummert, "Distributed dynamic channel assignment in wireless networks and its application to multi-channel ad hoc networks," in **Proceedings of the 7th International Conference on Systemics, Cybernetics and Informatics (SCI)**, Orlando, FL, USA, July 2003, vol. III, pp. 492–495.

[88] J. Zander, S.-L. Kim, M. Almgren, and O. Queseth, **Radio Resource Management for Wireless Networks**, Artech House, 2001.

[89] S. Ramanathan, "A unified framework and algorithm for (T/F/C)DMA channel assignment in wireless networks," in **Proceedings of the IEEE Conference on Computer Communications (INFOCOM)**, Kobe, Japan, Apr. 1997, pp. 900–907, Also appeared in ACM-Baltzer Journal of Wireless Networks, 5 (1999), pp. 81–94.

[90] G. Coulouris, J. Dollimore, and T. Kindberg, **Distributed Systems: Concepts and Design**, Addison Wesley, 2001.

[91] J. J. Garcia-Luna-Aceves and L. Bao, "Distributed dynamic channel access scheduling for ad hoc networks," **Journal of Parallel and Distributed Computing – Special Issue on Wireless and Mobile Ad Hoc Networking and Computing**, vol. 63, no. 1, pp. 3–14, Jan. 2003.

[92] X. Ma and E. L. Lloyd, "Evaluation of a distributed broadcast scheduling protocol for multihop radio networks," in **Proceedings of the IEEE Military Communications Conference (MILCOM)**, Washington D.C., Oct. 2001.

[93] E. L. Lloyd and S. Ramanathan, "Efficient distributed algorithms for channel assignment in multihop radio networks," **Journal of High Speed Networks**, , no. 2, pp. 405–428, 1993.

[94] J. J. Garcia-Luna-Aceves and J. Raju, "Distributed assignment of codes for multihop packet-radio networks," in **Proceedings of the IEEE Military Communications Conference (MILCOM)**, Monterey, Nov. 1997.

[95] T. H. Cormen, C. E. Leiserson, and R. L. Rivest, **Introduction to Algorithms**, The MIT electrical engineering and computer science series. MIT Press, McGraw-Hill, 1990.

[96] S. Prill, "Entwicklung und Implementierung eines verteilten Algorithmus zur dynamischen Kanalzuweisung in funkbasierten ad hoc Netzwerken," Diploma thesis, Communication Theory, University of Wuppertal, Germany, May 2002, (in German).

[97] A. Honarbacht and A. Kummert, "WSDP: Efficient, yet reliable, transmission of real-time sensor data over wireless networks," in **Wireless Sensor Networks, First European Workshop, EWSN 2004, Berlin, Germany, Proceedings**, H. Karl, A. Willig, and A. Wolisz, Eds., number 2920 in LNCS, pp. 60–76. Springer Verlag, Jan. 2004.

[98] R. E. Kalman, "A new approach to linear filtering and prediction problems," **Transactions of the ASME: Journal of Basic Engineering**, vol. 82, no. 1D, pp. 35–45, Mar. 1960.

[99] I. R. Petersen and A. V. Savkin, **Robust Kalman Filtering for Signals and Systems with Large Uncertainties**, Birkhäuser, 1999.

[100] S. M. Bozic, **Digital and Kalman Filtering**, Arnorld, 2nd edition, 1994.

[101] C. K. Chui and G. Chen, **Kalman Filtering with Real-Time Applications**, Springer Verlag, 2nd edition, 1991.

[102] A. Honarbacht, F. Boschen, A. Kummert, and N. Härle, "Synchronization of distributed simulations - a Kalman filter approach," in **Proceedings of the IEEE International Symposium on Circuits and Systems (ISCAS)**, Phoenix, AZ, USA, May 2002, vol. IV, pp. 469–472.

[103] A. Kummert and A. Honarbacht, "State estimation of motion models based on asynchronous and non-equidistant measurement updates," in **Proceedings of the 8th IEEE International Conference on Methods and Models in Automation and Robotics (MMAR)**, Szczecin, Poland, Sept. 2002, pp. 55–60.

[104] I. F. Akyildiz, W. Su, Y. Sankarasubramaniam, and E. Cayirci, "Wireless sensor networks: A survey," **The International Journal of Computer and Telecommunications Networking**, vol. 38, pp. 393–422, 2002.

[105] R. Intanagonwiwat, C. Govindan and D. Estrin, "Directed diffusion: A scalable and robust communication paradigm for sensor networks," in **Proceedings of the ACM/IEEE International Conference on Mobile Computing and Networking (MobiCom)**, Boston, MA, USA, Aug. 2000, pp. 56–67, ACM.

[106] A. Honarbacht and A. Kummert, "WWANS: The wireless wide area network simulator," in **Proceedings of the IASTED International Conference on Wireless and Optical Communications (WOC)**, N. C. Beaulieu and L. Hesselink, Eds., Banff, Alberta, Canada, July 2002, pp. 657–662, ACTA Press.

[107] A. Honarbacht and A. Kummert, "Wireless adhoc networks in the metropolitan/wide area: Concepts and first results," in **Advances in Communications and Software Technologies**, N. E. Mastorakis and V. V. Kluev, Eds., Electrical and Computer Engineering Series, pp. 121–126. WSEAS Press, 2002.

[108] A. Honarbacht and A. Kummert, "A new simulation model for the 802.11 DCF," in **Proceedings of the International Workshop on Multimedia Communications and Services (MCS)**, Kielce, Poland, Apr. 2003, pp. 71–76.

[109] X. Zeng, R. Bagrodia, and M. Gerla, "GloMoSim: A library for parallel simulation of large-scale wireless networks," in **Proceedings of the 12th Workshop on Parallel and Distributed Simulation**, Banff, Alberta, Canada, May 1998, pp. 154–161, ACM.

[110] R. Brown, "Calendar queues: A fast O(1) priority queue implementation for the simulation event set problem," **Communications of the ACM**, vol. 31, no. 10, pp. 1220–1227, Oct. 1988.

[111] H. T. Friis, "A note on a simple transmission formula," in **Proceedings of the IRE**, May 1946, vol. 41, pp. 254–256.

[112] J. Doble, **Introduction to Radio Propagation for Fixed and Mobile Communications**, Artech House, 1996.

[113] A. Schlüter, "Entwurf und Implementierung eines Protokolls zur relativen Positionsbestimmung in drahtlosen, selbst-organisierenden Kommunikationsnetzen," Diploma thesis, Communication Theory, University of Wuppertal, Germany, June 2004, (in German).

[114] P. Rauschert, A. Honarbacht, and A. Kummert, "Performance analysis of the IEEE 802.11 timer synchronization function when applied to multi-hop ad hoc networks," in **Proceedings of the International Workshop on Multimedia Communications and Services (MCS)**, Kraków, Poland, May 2004, pp. 65–70.

[115] W. Tuttlebee, Ed., **Software Defined Radio: Baseband Technology for 3G Handsets and Basestations**, John Wiley & Sons, 2004.

[116] S. Mirabbasi and K. Martin, "Classical and modern receiver architectures," **IEEE Communications Magazine**, vol. 38, no. 11, pp. 132–139, Nov. 2000.

[117] H. Harada and R. Prasad, **Simulation and Software Radio for Mobile Communications**, Artech House, 2002.

[118] W. Tuttlebee, Ed., **Software Defined Radio: Enabling Technologies**, John Wiley & Sons, 2002.

[119] J. B. Groe and L. E. Larson, **CDMA Mobile Radio Design**, Artech House, 2000.

[120] H. Tsurumi and Y. Suzuki, "Broadband RF stage architecture for software-defined radio in handheld terminal applications," **IEEE Communications Magazine**, vol. 37, no. 2, pp. 90–95, Feb. 1999.

[121] T. Hentschel and G. Fettweis, "The digital front end - bridge between RF and baseband processing," in **Software Defined Radio: Enabling Technologies**, W. Tuttlebee, Ed., pp. 151–198. John Wiley & Sons, 2002.

[122] R. E. Crochiere and L. R. Rabiner, **Multirate Digital Signal Processing**, Prentice Hall, 1983.

[123] M. Cummings and S. Haruyama, "FPGA in the software radio," **IEEE Communications Magazine**, vol. 37, no. 2, pp. 118–123, Feb. 1999.

[124] C. Schaefer, "Erstellung eines Daughter-Boards für eine Texas Instruments Entwicklungsplattform zwecks Weiterentwicklung eines Rapid-Prototyping-Systems für drahtlose Übertragungsstrecken," Diploma thesis, Communication Theory, University of Wuppertal, Germany, Apr. 2004, (in German).

[125] A. Honarbacht, M. Krips, and A. Kummert, "Hardware concept for a multi-channel wireless ad hoc transceiver," in **Proceedings of the International Conference on Computer Communication and Control Technologies (CCCT)**, Orlando, FL, USA, July 2003, vol. II, pp. 219–224.

[126] M. Nahr, "Beiträge zu Entwurf, Implementierung und Verifizierung von Hardwarekomponenten einer W-CDMA Demonstratorstrecke," Diploma thesis, Communication Theory, University of Wuppertal, Germany, May 2004, (in German).

[127] M. Pradier, "Filterdesign für den Prototyp eines Senders für drahtlose Netzwerke der nächsten Generation und Akquisition des digitalen Zwischenfrequenzsignals," Diploma thesis, Communication Theory, University of Wuppertal, Germany, Sept. 2003, (in German).

[128] S. M. Alamouti, "A simple transmit diversity technique for wireless communications," **IEEE Journal on Selected Areas in Communications**, vol. 16, no. 8, pp. 1451–1458, Oct. 1998.

[129] P. Rauschert, A. Honarbacht, and A. Kummert, "R&D platform for MAN scale wireless ad hoc networks - towards a hardware demonstrator for 4G+ systems," in **Proceedings of the IEEE International Symposium on Circuits and Systems (ISCAS)**, Vancouver, British Columbia, Canada, May 2004, vol. IV, pp. 433–436.